Plants of the New World
The First 150 Years

The town of Secota, in present-day North Carolina, showing fields of tobacco and maize. From J. Hariot, *A Briefe and True Report of the New Found Land of Virginia.* Frankfurt am Main, 1590. (Catalogue item number 36.)

Plants of the New World
The First 150 Years

*An Exhibition of some Books which made known
the New World to Europe*

Elizabeth A. Shaw
Harvard University Herbaria

HARVARD COLLEGE LIBRARY
CAMBRIDGE, MASSACHUSETTS
1992

ON THE FRONT COVER:

"Maraco Indica," a species of *Passiflora*, the passion flower. From Vallet,
Le jardin du roy tres Chrestien Loys XIII. Paris, 1623. (Catalogue item number 40.)

ON THE BACK COVER:

Brazilnut tree and fruit, *Bertholletia excelsa*. From W. Pies and G. Marcgrave,
Historia naturalis Brasiliae. Leyden and Amsterdam, 1648. (Catalogue item number 16.)

DESIGN: GINO LEE

PRINTED AT THE OFFICE OF THE UNIVERSITY PUBLISHER

HARVARD UNIVERSITY

THE TYPEFACE IS MINION

AND THE PAPER MOHAWK SUPERFINE

The Exhibition

This exhibition in The Houghton Library of Harvard University is one of a series of events presented under the sponsorship of the Columbian Quincentenary Committee of Harvard University to mark the five hundredth anniversary of the first voyage of Christopher Columbus to the Americas.

A subvention generously given by the Real Colegio Complutense, a joint project of the Complutense University of Madrid and of Harvard University, undertaken to promote and to encourage scholarly work, made possible the production of this catalogue. Other significant contributions to this event have been made by the Office of the President of Harvard University, by the University's Committee on Latin American and Iberian Studies, and by the Friends of the Harvard College Library.

Sunflower, *Helianthus annuus.* From Camerarius, *Symbolorum et emblematum ex re herbaria...collecta.* Nuremburg, 1590. (Catalogue item number 41.)

Acknowledgements

This exhibition and catalogue have grown from a long-time interest in the early literature of natural history and in paths of information flow in that subject. For this opportunity to express these interests I am deeply grateful to the Columbian Quincentenary Committee of Harvard University and its chairman, Professor John Womack, Jr.; to Professor Otto Solbrig; and to The Houghton Library.

Many colleagues and friends have listened to, and discussed with me, the views expressed here and I am grateful for their patience and willingness to exchange ideas. I wish to extend special thanks to Professor Donald H. Pfister, Director of the Harvard University Herbaria, for encouragement and support during this work, and to A. Hunter Dupree, historian of American science, for his advice and enthusiasm. I wish, too, to thank Judith A. Warnement, Librarian of the Botany Libraries of Harvard University, for her cooperation and interest in using in the exhibition books from the libraries in her charge.

My particular thanks go to Roger E. Stoddard, Curator of Rare Books in the Harvard College Library; to James J. Lewis, Curator of the Reading Room in Houghton Library, and to the staff of the Reading Room for their patience and many courtesies.

Elizabeth A. Shaw
Harvard University Herbaria

Preface

The twin themes of Harvard University's Columbian Quincentenary celebration are discovery and exchange. Discovery, because the voyages made by Columbus resulted in the discovery of each other by peoples who had been isolated before historical memory. Exchange, because discovery was followed by a transatlantic exchange of technologies and artifacts. Agricultural information was one of the principal exchanges. European agriculture in the fifteenth century was based on the planting of fine-grained cereals—primarily barley, wheat, rye, and oats—in fields cultivated by plows drawn by oxen or horses and sown by broadcast of the seeds. American agriculture was based on the use of large-seeded plants, such as maize, squashes, and beans, planted individually with digging sticks. Further south, in the Andes, an agriculture depending heavily upon rootcrops (potatoes, manioc, oca, and achira) had developed. Exchange helped to produce the mixed agriculture practiced today on both sides of the Atlantic.

Columbus took back with him some of the new American plants he encountered, such as tobacco, pineapples, and maize, for exhibition at the Spanish court. There they made a great impression. On his second voyage he brought to America seeds and cuttings of the plants which Europeans valued (wheat, barley, sugar cane, and grapes) and they soon were growing in Hispaniola, Cuba, and on the Mesoamerican mainland, next to the native crops. In the next one hundred years maize, potatoes, tomatoes, and tobacco, invaded Europe and soon became so entrenched in European cuisine and culture that we often forget that they are newcomers.

Seeds alone are not sufficient to grow a crop successfully. Knowledge about sowing times; a plant's requirements for soil, water, and temperature; when and how to harvest; and, above all, what to harvest, is needed. There are many stories of people eating the poisonous fruits of the potato rather than the nourishing tubers.

Information on the new world, including its agricultural practices, was disseminated through the writings of explorers and visitors. The voyages of Columbus coincided with the spread of printed books, made affordable by Gutenberg's invention half a century earlier. Accounts of travelers were very popular and must have fascinated Europeans—most of whom never travelled beyond their local area.

In this exhibition we wish to point out how information on America and its technology, especially botanical and agronomic information, reached Europe. The focus is on information as much as on the plants. For this purpose books written by many kinds of people have been selected: explorers, administrators, soldiers, missionaries, and others. They came from many parts of Europe, and they visited different parts of America. People brought back not only edible plants, but also those of medicinal and horticultural value, as documented in herbals and medical treatises of the time.

Arrangement of an exhibit such as this one takes a great deal of preparation. The original plans for all of the exhibits were drawn up by the University's Quincentenary Committee in 1989. But it was the vision and the work of Dr. Elizabeth Shaw of the Gray Herbarium that made it possible. For more than a year she carefully and painstakingly studied the existing documentation on the transfer of botanical knowledge during the first hundred years after the discovery in order to select representative and informative works. She was aided in this work and in the production of the catalogue by Roger Stoddard, Curator of Rare Books in the Harvard College Library.

This exhibition is made possible through the financial assistance of the Real Colegio Complutense and the Office of the President of Harvard University. We would like to thank Professor Evon Vogt, who initiated the idea of these celebrations; Professor John Womack, Jr., who ably chaired the Quincentenary Committee; Dr. Enrique Alonso of the Real Colegio Complutense for his enthusiasm and support; and Dr. Anne Normann, executive director of the University's Committee on Latin American and Iberian Studies, who ably coordinated the entire effort, and has steered it happily through the unavoidable reefs of minor and major crises.

Otto T. Solbrig
Bussey Professor of Biology

Introduction

Columbus

Late in the evening of the 11th October, 1492, Christopher Columbus, watching and hoping to see the Indies, stood on the sterncastle of the *Santa Maria* and thought that in the distance there was a light. As told in Bartolomé de las Casas' abstract of Columbus' own journal of the first voyage, a few hours later Juan Rodríquez Bermejo, standing watch on the *Pinta,* called out that he saw land. At first light the anxious watchers on the three small ships saw other interested watchers—people standing on the beach of an island in the Bahamas. Columbus and the captains of the *Pinta* and the *Niña* went ashore where they broke out the royal standard and other flags.

The reader familiar with narratives of discovery might expect that the next event told would be the formal possession of the land for Ferdinand and Isabella, and, indeed, that was the next act. But there comes first in de las Casas' narrative this sentence—"once ashore they saw very green trees, many streams, and fruits of many kinds."

"Green trees...and fruits of many kinds"—here is the first hint of the biological wealth of this world new to Europeans, and the first hint of events which would change the worlds known to both sets of watchers on that morning.

Columbus had reached this land with its green trees by sailing westward to find the east, drawn by the knowledge that in the east—wherever that might be—lay lands from which came to Europe desirable, and costly, plant products—spices and medicines. Europeans knew this through the realities of the marketplace and literate Europeans could know something of them by reading Pliny's *Naturalis historia,* printed in 1469, or by reading works on medicine or by reading what Marco Polo had written about those parts of the east which he had visited or which he knew by report. Many of his readers might not believe what they read, but for those who did, there were wonders, including reports of gold mines and palaces roofed with gold in Cipangu.

How Columbus came to believe in his Empresa de las Indias we do not know—whether he slowly stitched together pieces of information, many of them inaccurate, about the extent of the oceans and the circumference of the earth, or whether it came to him in a flash of intuitive insight. We do know that Columbus had personal experience of Portugal's slow push down the western coast of Africa, for in the 1480's he had gone as far as São Jorge da Minas in Ghana.

We know, too, that Columbus owned and read books. Some of them, including his own copy of Marco Polo, survive in the Biblioteca Columbina in Seville. He used his books, seizing any statements which seemed to support his contention that the shortest route to the east lay westward. Of great importance to Columbus, evidenced by the nearly nine hundred annotations made in his copy, was *Imago Mundi* of Pierre d'Ailly, written about 1410 and printed with some of his minor works in 1480-1483. However, Cardinal d'Ailly's picture of the world was a mixture of fantasy and fact. D'Ailly quoted Aristotle's comment in *De caelo* that the ocean joins the Pillars of Hercules and India—true enough—but he greatly overestimated the longitudinal extent of the world's landmass, assigning to the land 225 degrees of longitude and only 135 degrees to the ocean.

Columbus' estimate, based on a combination of accurate and inaccurate information (and some arbitrary assumptions about distances), of the distance between the Canary Islands and Japan was 2400 nautical miles. In fact, that distance is more than 10,000 miles. Columbus underlined and annotated and accepted both facts and fictions, and set out westward across the ocean which would be the shortest path to the east. The lines from the *Medea* of Seneca quoted on the title page of this catalogue, with their prophecy that the oceans would be the highways by which the world would become one, were favorites of Columbus. His son Hernando wrote in his father's copy of Seneca, "This prophecy was fulfilled by my father, the Admiral, in the year 1492."

Of all of the consequences of that encounter between Europeans, and the people already in the "new world" which lay between Europe and the eastern lands which Columbus hoped to reach, that which has had the greatest impact upon the descendants of those who lived at the time of the encounter was the "exchange" of living systems, of biological information, between the known and the newly-found lands.

The "Columbian exchange," is man's post-Columbian transportation, knowingly or not, of animals, of plants, and of disease-causing organisms from landmass to landmass. Europeans brought cattle, swine, sheep, chickens, goats—those omnivorous beasts—and horses (formerly here, but by the time of the encounter long extinct in America). These are all species brought by Columbus on his second voyage of 1493 into a world which presented to them little in the way of predators or pathogens. Brought, too, on the second voyage were some of the familiar cereals which were the staples of old world agriculture and the staple vegetables (carrots, lettuce, and peas) of Spain. Soon brought as well were figs, dates, several species of citrus, and sugar cane.

Thirty years later Gonzalo Fernández de Oviedo, gone to the new world in 1514 as supervisor of goldmining and smelting, returned to Spain and wrote for Charles V a "summary of the natural and general history of the West Indies." Oviedo was a sensible and careful observer and I think that we can accept as the expression of accurate observation such statements as these: "plants native to Spain that have been transplanted and cultivated there grow better and in larger quantity than in any part of Europe" and "cows have multiplied at such a rate that many cattle kings have more than a thousand or two thousand head, and there are quite a number who have up to three or four thousand head." Within thirty years or less these immigrants had made good. Europeans brought, too, their familiar diseases, smallpox and measles and others, which swept through native peoples leaving devastating depopulation.

From the Americas have come only a few domesticated animals, notably the guinea pig and the turkey, each having, in English at least, a geographically derived name which is wildly off the mark. Diseases—most likely the new world sent back syphilis, but it is still not known with certainty where that disease originated.

Not much of an exchange, one might say. Columbus sailed westward to reach the east from which came the plant products so much wanted in fifteenth-century Europe—the cinnamon, cloves, nutmegs, and pepper, greatest of them all. These he did not find for they do not occur in the Americas. But Columbus and his successors did find plants which have changed the world, in small ways and in great—such a diverse array as tobacco, cacao, peanuts, rubber, tomatoes, chilies, quinine, and the two great carbohydrate staples of the new world, maize and the potato, which have built populations and changed demographics.

These have been the true riches of the Americas—not the gold washed from streams and made into the objects found by Cortés and Pizarro, not the silver taken from Potosí, both of them finite in quantity and not renewable in human timespans. The plants are renewable resources—perhaps the archetypic renewable resources. Plants of the Americas have brought new foods, carbohydrate staples, vitamin-rich "vegetables" and fruits which have altered agricultural practices and dietary patterns in much of the world; they have provided compounds such as quinine which changed life in much of the tropics; they have provided the raw materials for many industrial processes. If we can refrain from damaging in an irreparable way the world we share with so many others, there will remain the potential for many other such discoveries.

No doubt man always has found attractive many of the scents and flavors of plants. Plants show an amazing array of biosynthetic pathways by which are built a great diversity of compounds which we find pleasing in odor and taste or which are pharmacologically active; some are all of these. Thousands of species synthesize such substances; a few of them have been used for millenia—on our bodies and clothing as perfumes, to flavor our food, to heal ourselves, and to make offerings to our gods. Although in modern English the word "spice" has a restricted meaning and refers only to certain flavorings, the word has been used in the past in a much broader sense, and I shall so use it here.

Pliny, writing about A.D. 70, recognized three kinds of spices; "aromata"—the fragrant substances; "thymiamata"—the incenses burnt as offerings to the gods; and "condimenta"—used as seasonings—modern spices. Plant materials used as simples or in compound medicines he set aside from these.

The Graeco-Roman world had at hand many species used as condiments and in medicines. Growing much for the collecting were plants used still today—basil, parsley, mustard, mints, and others. But the most desired spices came to Rome by trade routes which originated to the south or far to the east. The Hellenic world had known for several centuries of some of the products brought from the east. Theophrastus, writing about B.C. 300, listed nineteen plants used in perfumes, and said of them, "Some of them grow in many places, but the most excellent and most fragrant all come from Asia and sunny regions. From Europe itself comes none of them except the Iris [orris-root]." As Rome became hellenized, knowledge of the east and its products was absorbed so that by the time of Augustus, commerce between Rome and "Asia" was well-developed and there was increasing demand in the west for products of the east.

The Roman end of the commercial exchange consisted of manufactured goods such as glass and textiles, and a few natural products, notably coral, but these were not of value enough to balance the cost of the imports—the pepper, the cinnamon and cassia, the malabathrum, the myrrh and frankincense (each of these valued as highly as gold), and lesser spices. The difference was made up in Roman gold; the problem of trade balances with the east is not new. Pliny pointed out that the imbalance drained specie away from the empire and complained that Rome lost at least 100 million sesterces each year—"that is the sum which our luxuries and our women cost us."

The greatest of the luxuries was pepper, *Piper longum* from northern India and *Piper nigrum* from the south. Pepper seems to have come late to Rome—Horace noted it about B.C. 15 as a rarity, wrapped in small twists of paper—but one hundred years later Domitian had built the *horrea piperataria,* warehouses for pepper, and by A.D. 408 when Alaric lay siege to Rome, he demanded, and received, 3000 pounds of pepper as part of the city's ransom. Pepper never lost its popularity. One thousand years after Alaric, pepper accounted for sixty percent of the value of spices coming into Venice, then the center of the trade.

Many writers such as Dioscorides, Pliny's near contemporary, whose *De materia medica* was the basis of pharmacy for 1500 years, noted the origins of spices. Indeed, Rome was intrigued by origins, but in general the geographical origin of products which came from great distances was poorly known. The stated provenance often was "India," and that was correct enough for pepper and some of the lesser spices. Several were said to come from "Arabia felix"—the Hadramaut in southern Arabia. Some did, indeed, originate there; myrrh and frankincense, for example, are resins produced by small trees native to Yemen and to Somalia in east Africa. Cinnamon and cassia were also attributed to *Arabia felix,* but their true point of origin was southeastern China; many of the regions thought to be producers of goods imported into Rome were commercial centers to which came trade routes that originated great distances away. Biological provenance was equally mysterious. Pliny remarked that pepper grew on trees "like our junipers;" not so of course—he confused the small, berry-like cones of *Junipe-rus communis* with the fruits, true berries, borne by the climbing vine, *Piper nigrum.*

Alaric's demand for pepper marks the beginning of the end to Roman authority in the west, but even as that authority collapsed, imports from the east continued. Patterns of trade shifted; prices rose and the sources of the plant materials became comparatively more remote, although western Europe never was completely shut off from eastern goods. Still, to the average European of A.D. 1000 the east was quite unknown and its products might as well have fallen from heaven.

From the end of the eleventh century western Europe went crusading and many "Franks" became familiar with the immediate near-east, although Jerusalem was the *terminus ad quem* and there was no reason for a devout Christian to go beyond. During the thirteenth century the picture changed. Early in the century the Tatars moved from Mongolia into China and by the middle of the century they had built the largest empire then known, extending from the China Sea to eastern Europe.

Their relatively stable administration of this huge area made possible relatively safe travel. During the century-long open window of the "Mongol Peace" a stream of European clerics and merchants of whom the Polos are only the best known, travelled to China and back, most often by a land route through Samarkand and Kashgar into northwestern China. This window closed in the middle of the fourteenth century—the result of westward spread of bubonic plague, of conversion of the smaller Mongol khanates in Central and Western Asia to Islam, and of the overthrow of the khanate in China by the Mings in 1368.

Disruption of the caravan routes across central Asia left open a southern route which was a far more complicated series of steps of trade, with goods being shipped from one commercial node to the next. Goods from as far east as China and Indonesia were brought via Malacca to Sri Lanka and the Malabar coast of southwestern India, and were then transshipped by Persian and Arab traders to Hormuz on the Iranian coast, to Aden at the mouth of the Red Sea, or even to ports on the Horn of Africa. From these ports eastern imports ended up at Acre, Beirut, and, especially, at Alexandria where ownership was transferred from Muslim to Christian hands. In this trade the most important spices were ginger and cinnamon from southeastern China; cloves from the Molucca Islands, known until the eighteenth century as the "Spice Islands;" and, of course, pepper. The products eventually reached European markets, but their prices reflected the adding on of costs at each point of transshipment, including heavy duties imposed at the eastern Mediterranean ports.

Trade in spices, even in a broad sense, was a small part of the general pattern of commerce in the eastern Mediterranean and the Levant, but the spices were the high-value items and they were unique; there was nothing which could replace them. There was only a limited range of European goods which could be sold profitably in the near east, and the high-value goods had usually to be paid for with gold and silver which were both in short supply in fifteenth-century Europe.

The shortage of monetary metals emphasized to Europeans the importance of trade with north Africa, especially that based on European goods such as copper, textiles, and Venetian glass beads carried to the south by trans-Saharan caravans. They returned with slaves, some ivory and fancy leathers, but, most importantly, with gold which came from unknown sources south of the desert.

By the middle of the fifteenth century the Iberian push into the eastern Atlantic and down the coast of Africa, led by Portugal, was well-advanced. There were many stated reasons—the wish to find other Christian rulers, to convert non-Christians, to do noble deeds, and to discover the unknown—and the underlying economic necessity of finding gold. The "program" was successful enough that in 1481 Portugal built a fort and factory at Elmina on the coast of Ghana. Columbus seems to have made at least one voyage there.

By 1490 there had developed on the African coast an economically viable trade, based upon gold, slaves, and the pungent malagueta pepper, *Aframomum malagueta,* not as finely flavored as *Piper nigrum* from India, but much less expensive. Columbus had seen the development of this new commerce and the lure of sailing a comparatively short distance westward to reach the "real" source of spices and of the gold reported by Marco Polo, as he had long believed could be done, must have been irresistible.

THE EXHIBITION

None of the plant products—the spices—which Columbus hoped to find occurred in the new lands which lay across his westward route to Asia. This was biogeographical reality and what Columbus and those who came after him did find was indeed a biologically new and diverse world.

The aim of this exhibition is to present a sampling of the books which made known to Europeans the plants and the plant products of the new world during the 150 or so years after the encounter. This offers a well-defined time frame, for the middle years of the seventeenth century are the line of demarcation between the end of the late medieval understanding of the biological world and the start of early-modern biology. At the end of this period appear the first works which offer broad surveys of particular areas, notably of Mexico and of part of Brazil. With regard to the plants and the ways in which they were used, one sees the beginnings of plantation agriculture, or the growing of plants as monocultured commodities, to bring wealth to European countries developing as colonial powers.

The greater number of books in this exhibition are either works of travel or herbals, both classes quite broadly delimited. The literature of travel was written by those who had been in America and could speak with first-hand knowledge of what they had seen and experienced there, or by men who were one step removed from American reality by reporting the news brought home to Europe. Of these latter the first, and most influential, was Peter Martyr, cleric and intimate at the Spanish court.

None of these early authors was a botanist. It would be eighty years after the first voyage of Columbus that a physician, Francisco Hernández, would go to the new world for the purpose of reporting upon American plants, and his work would remain unprinted until the middle of the seventeenth century. A few recent writers have pointed to the fact that Columbus had with him no botanist—as if that were some grievous fault—without understanding that in Europe in 1492 there was not one person who could have gone with Columbus and told him anything more about the biological world around him than he recorded for himself. European botany lay then in the hands of a few humanist scholars who were resolving textual problems in Pliny's *Naturalis historia.*

The books which brought home to Europe first-hand information about new world plants were written, then, not by botanists, but by administrators, adventurers, explorers, missionaries, and soldiers. They were a diverse lot of people who observed with interest and intelligence the new world and who wrote, for a variety of reasons, about what they saw. These works presented to Europeans a richness of information about the natural products of the new world and about their use by American peoples.

By the first years of the sixteenth century we see, on the one hand, the beginning of the flow of information about the natural history of a world entirely new to Europe; and on the other, the concern of humanist scholars not with the new, but with first-century texts. The two works which transmitted to Renaissance Europe information about plants were the *Naturalis historia* and, more important, *De materia medica,* written in Greek by Dioscorides about A.D. 60. For 1500 years this work, transmitted to western Europe in Latin translation, variously modified, abridged or augmented, would be the definitive work on pharmacy. In the middle of the fifteenth century, Greek manuscripts of Dioscorides were brought from Constantinople to Italy, and the true text of *De materia medica* was printed in Latin in 1478 and in Greek in 1499. For much of the next century the concern of European "botany" was the attempt to correlate the plants described by Dioscorides and, therefore, prescribed by physicians, with those actually sold by apothecaries. The great herbals of the sixteenth century are basically commentaries on Dioscorides.

Gradually, however, mention of American plants comes into these works. The first is guaiacum, the wood of *Guaiacum officinale* from the West Indies, brought into European medicine as a "cure" for syphilis which, in a highly infectious and often fatal form, ravaged Europe during the first fifty years after the encounter. One thread to be followed in the exhibition is the mention of more new world plants in the herbals and the slow evolution of the herbals from Dioscoridean commentaries to more general works on plants.

New world plants were of interest for other reasons as well. The interest of Renaissance Europe in the rare and the marvelous was piqued by such oddities as cacti and even by maize which may have been grown as a novelty as early as the 1520's. By 1560, in Conrad Gesner's listing of plants grown in German gardens, we find several Americans—maize, marigolds, tomatoes, prickly pears, and others. By the beginning of the next century many more new world plants, especially those introduced to the royal gardens in Paris from French Canada, were being grown.

Knowledge of these new plants gradually spread from those interested in them as medical drugs to a more general awareness of them. By the end of the sixteenth century Shakespeare could refer in *The Merry Wives of Windsor* to potatoes—sweet potatoes, in fact—and expect that many in his audience would appreciate the reference.

Last, and eventually the most important, there appears the start of the movement of plants, from food to weeds, around the world. If any one thing fulfilled the prophecy inherent in the lines from Seneca's *Medea* which were favorites of Columbus and which appear on our title page, it was the movement of organisms, of biological information, along the routes which were opened as a consequence of the trans-oceanic explorations, both west and east, at the end of the fifteenth century. Plants are especially easy to carry. Unlike Noah's Ark, which had to carry pairs of animals that were sexually mature or would become so, all that is required, depending upon the species, is a handful of seeds or a few roots, tubers, or stem cuttings. A single ear of maize will provide some hundreds of genetically distinct individuals. In some species a single seed can found a new population.

The easily-carried and grown food plants of the new world, maize, sweet potatoes, and cassava, were transported around the world almost from the time of their discovery by Europeans. New world food plants were taken to India and beyond by Portuguese seamen at the start of the sixteenth century and some of them were in China by the middle of the century. One hundred years later Michael Boym, a Jesuit missionary, published the first western account of plants of China. Five of the eighteen plants he mentions are Americans.

It is more than chance that the opening of ocean routes with the consequent binding together of the great landmasses came only forty years after the development in Europe of printing with movable type. There had been earlier contacts between the Americas and the old world, but they had been forgotten or faded into myth. If the oceanic routes were highways over which information was exchanged, so were the books which could be printed in their many hundreds and distributed as carriers of information to be read and shared.

So we wish in this exhibition to observe the five-hundredth anniversary of the first voyage of Columbus and to celebrate the books through which a part of the new world was made known in Europe. It was a huge and varied world and I have tried to reflect this diversity in the parts of the new world described, from Canada to southern Brazil; in the authors; and in the books themselves, written in several vernaculars and in Latin, and printed in ten European countries.

A last point to note is the value of the informational content of the books for the modern world. For some who work in the sciences it is all too easy to ignore the past or to reject it as of no possible interest, but there is in these books a great wealth of information concerning the reality of America before it was much touched by European influences. It is against such a background that we can seek to avoid repeating the errors of the past and better to understand the present.

NOTES ON SOME OF THE PLANTS

The first news of new world plants reached Europe by *Epistola de insulis noviter repertis,* the "Columbus letter," many times printed in 1493 and the following years. Columbus had to "prove" to Ferdinand and Isabella that he had indeed reached the Indies and one way to do this was to list the spices which he had found. Among the plant products reported by Columbus were "spice," by which he meant pepper; cinnamon, mastic, aloe, and rhubarb, although none of these items was the valuable commodity which reached Europe from the "East."

The plant product which Columbus most hoped to find was pepper, the archetypic spice of medieval Europe. The peppercorns which reached Europe were the fruit of *Piper nigrum,* grown in southern India. Black pepper was the fermented and dried berries, while white pepper was the same thing, but with the outer layers of the fruit removed. Columbus found nothing like this, but he did find the pungently flavored fruits of *Capsicum annuum* (pimento, paprika, cayenne and bell peppers) and *C. frutescens,* the tabasco pepper, which were commonly used by the Arawak people.

The report of cinnamon in the letter seems to have been unfounded optimism. Columbus noted in his journal for 4 November, 1492, that Martín Alonzo Pinzón had brought cinnamon to him, and on 6 November that the Arawak told him that pepper and cinnamon, of which Columbus had brought samples, grew not "here," but "nearby." After these two entries cinnamon is not again mentioned. What Pinzón brought may have been dried and rolled leaves of tobacco which might have resembled the dried and rolled quills of bark which are the product "cinnamon."

Mastic was the one item which Columbus knew at first hand as a plant, rather than as a product. It is the resin of *Pistachia lentiscus,* a small tree native around the shores of the Mediterranean. In the fifteenth century, production was centered in the island of Chios, held by a Genoese association, the Maona or Giustiniani, which strictly controlled gathering and export of the resin; Columbus had been to Chios and seen the operation for himself. Mastic has now no commercial value but in the fifteenth and sixteenth centuries it was much used as an ingredient of compound medicines. The "mastic" trees noted by Columbus on Cuba and Hispaniola were *Bursera simaruba,* the West Indian gumbo limbo.

On 21 October Columbus recorded that on Crooked Island in the Bahamas he had seen "aloes" and decided to load ten *quintals* of it, nearly one thousand pounds. On the next day, "I made them gather as much of the aloes as could be found." Aloe as an article of commerce at that time was made from the pulpy tissue of leaves of *Aloe barbadensis* (in spite of its name, not a native of America) and related species, most of it coming from the Horn of Africa and the island of Socotra. The liquid in the pulp was allowed to drain off and was then evaporated, ideally to a solid residue, which was a valuable medicinal product used as a purgative.

Columbus used the word "lignaloe" which suggests that he might have confused the medicinal aloe with a quite different substance, "lignum aloes" or "aloe wood," the resinous wood of *Aquilaria agallocha,* an Indian tree once used for incense. The plants which he saw in the West Indies were species of *Agave,* which superficially look much like *Aloe,* although they are much larger. One thousand pounds of fresh *Agave* leaves on board ship would have quickly turned into a most unpleasant mess and the load was probably soon jettisoned.

A last product mentioned by Columbus in the letter and in his journal as being of value is rhubarb. This was not modern culinary rhubarb, *Rheum rhaponticum,* but other related species, the roots of which had been part of *materia medica* since Dioscorides prescribed "rhubarb," noting that the drug came from beyond the Bosphorus. In fact, the identity and origin of medicinal rhubarb would not be known with certainty until the early twentieth century, when it was shown that the product imported to Europe came from two Chinese species.

In medieval Europe rhubarb was one of the most expensive drugs, used as an effective, yet gentle, purgative. We may wonder that of the products mentioned by Columbus as being of value were two purgatives. They were used as are modern laxatives, but there was more to their use than that. European medicine was dominated by a Galenic view of the body and its health. Galen (ca. A.D. 130-200) presented the world and, on a lesser scale, the body, as composed of four elements, earth, air, fire, and water; each element was described by any two of four "qualities," heat, cold, wetness and dryness. In the body the elements and their qualifiers appeared as four humors—melancholy or black bile, choler or yellow bile, blood, and phlegm. In a very complex way the humors made up the body and perfect health depended upon a perfect balance of the humors; any disruption of this balance could cause disease. According to this scheme purgatives were of great value, for the physician could use them to restore balance among the humors or to evacuate those which were imperfect or were too abundantly present.

Brazilwood.   The first American species to become an article of commerce was brazil, *Caesalpinia echinata* and related species, which Columbus found on his third voyage which reached South America. Many species of trees sequester in the heartwood pigmented substances and some of these have long been used as dyes. The original brazil was *Caesalpinia sappan,* a leguminous tree native to southeastern Asia, which had for several centuries been imported into Europe. From the chipped heartwood was obtained a bright red dye, in color something like embers glowing in a brazier, and it has been suggested that "brazil" and "brazier" are etymologically related.

To return from the new world with a load of brazilwood was an easy way to make a quick profit and ships of several countries, especially of France, visited South America from early in the sixteenth century. The name of the dye was transferred to the wood, to the tree, and eventually to the region itself. Species of *Peltophorum* and of *Haematoxylon,* both members of the legume family, were also used as dye sources and at various times and places, any one of these could be called "brazilwood."

Haematoxylon campechianum, native to southern Mexico, was imported into England as "peach" or "campeachy" wood, after Campeche in Yucatán, whence it was exported. This, or a related species, was also known as "logwood" because it reached European markets as trimmed logs. In 1581 its use in England was forbidden, in part because the color, when used without a mordant, was fugitive; and, in part, because logwood competed in English markets with locally grown woad, *Isatis tinctoria.* The act which forbade its use was repealed eighty years later, but other dyes had nearly replaced the woods. The original brazilwood, *Caesalpinia echinata,* is no longer in commerce, but small quantities of species of *Haematoxylon* are still imported. The dye from the heartwood is used in dyeing leathers and furs and is the source of a biological stain used in plant histology.

Guaiacum and sassafras.   The first new world species to enter European medicine was *Guaiacum officinale,* a small tree native from southern Florida through the West Indies. First imported about 1510, guaic, guaiacum, or lignum sanctum was highly regarded as a cure for syphilis. The question of the origin of syphilis, whether the disease came from the new world or not, is not yet unequivocally answered, although there is an extensive literature on the subject. However, contemporary writers reported it as a new and virulent disease which appeared in 1494 or 1495 and swept—raged, rather—through Europe in a way we hardly can imagine.

It was generally accepted that syphilis first appeared in Italy and was carried across the Alps by the troops of Charles VIII of France who had invaded the peninsula in 1494. As the troops, many of them mercenaries, dispersed, so did syphilis spread. Sebastian Brant in 1496 reported that the disease was already in Germany and Thrace. All of the early reports agree that syphilis was highly contagious, that it caused severe pain, and that it frequently was fatal. Indeed, during its first forty to fifty years in Europe syphilis was far more severe than it was by the end of the sixteenth century and later.

One of the first symptoms was a rash and this was usually treated with mercury-containing ointments which were already in use in Europe to treat skin diseases. It was a case of the "cure" being nearly as bad as the disease, for the use of mercury caused loss of teeth and hair, paralysis, and, often, serious mental disorder. Guaiacum was first reported as being used in the West Indies to treat venereal diseases. Assuming that a new world plant might cure a new world disease, guaiacum was tried in Europe. Two sufferers who believed themselves helped published on guaiacum, Leonard Schmaus in 1518 and Ulrich von Hutten in 1519. Hutten's book was quickly reprinted and soon translated and became widely known.

The banking house of the Fugger family saw here a chance for profit and obtained from the Emperor Charles V a monopoly on importation and sale of guaiacum. Clever promotion by the Fugger cartel made guaiacum a valuable commodity during the 1520's and 1530's, but it, in fact, had no effect upon syphilis. The medicinal product was a drink made by steeping chips and shavings of guaiacum in water. The patient then drank as much of this as possible, was wrapped in blankets, and put to bed in a warm room. At best, the extract was a sudorific and the patient may have found comfort in thinking that he or she was sweating out the disease.

About the time that it was generally realized that guaiacum had no real medicinal properties, another new world wood, sassafras, became known in Europe. *Sassafras albidum* was brought from Florida in the 1560's and was first reported in Europe in 1571 by Monardes. It became the next "cure" for syphilis and other diseases and ailments—for a time a near-panacea—and sassafras quickly became a commodity. Vessels calling along the North American coast from southern New England to Florida had a good chance of finding sassafras in sufficient quantities to cut and load profitably. It certainly played a part in attracting English vessels to North America.

Sassafras, too, fell from favor and is now used only to give an attractive scent to soaps. It has still some little use as a folk remedy or tonic in the form of a tea, but sassafras has been shown to contain carcinogens and should not be consumed. *Guaiacum officinale* and the related *G. sanctum* are still articles of commerce. The wood, which has a high resin content, is hard and dense and is used in wooden machinery in situations where naturally lubricated bearings and bushings are of value. It is also used in turnery to make bowls, goblets, spoons and forks.

Cotton. On his first voyage Columbus found cotton, familiar to him from the Mediterranean, as spun thread and as fabrics. The fibers from which the thread was spun are single-celled hairs which grow from the seed-coat of species of *Gossypium,* a pan-tropical genus of shrubs and small trees. All species of *Gossypium* have very short seed hairs, but four of them have, in addition, longer ones, and these are the cultivated cottons.

Two Asian species, *Gossypium herbaceum* and *G. arboreum,* have been used for millennia, for cotton fabrics dated at about B.C. 3000 were found at Mohenjo-Daro in southern Pakistan. These short-stapled cottons were used in Roman times to produce fine and expensive fabrics. In America cottons dated ca. B.C. 2500 are known from Peru, made from *Gossypium barbadense* which, in spite of the name, is native to western South America. The cotton seen by Columbus was likely a race of *G. hirsutum,* native around the Gulf coast and from Cuba through southern Brazil. Evidence from chromosome number and size indicates that the American fabric cottons are derived from crosses between American and old world species. The simplest explanation is that old world species were carried across the Atlantic by natural means, most likely by drifting seeds. Nearly all of the cotton grown today is "upland" cotton, *G. hirsutum.* The very long-stapled (2.5–6 cm) "sea-island," "Egyptian," and "Pima" cottons are *Gossypium barbadense.*

Cassava. On the first voyage, Columbus tasted cassava, sweet potatoes, and maize, three of the American staple sources of carbohydrate. The fourth, the Andean potato, would be found in Peru forty years later.

The staple food of the islands and the northern and eastern coasts of South America was *Manihot esculenta,* a shrubby member of the *Euphorbiaceae,* the family to which belong Pará rubber and the poinsettia. In the islands it was known as "yuca," not to be confused with *Yucca,* a very different plant. Further south, in Brazil, the name was "mandioca" or something similar, which is anglicized to manioc. "Cassava" meant originally the flat bread made from the roots, but this is now the name used in commerce and in agronomy for the crop.

Known only as a cultivated plant, cassava is grown for its tuberous, starch-filled roots which can be two or more feet long and weigh several pounds. It is normally propagated by stem cuttings and makes, therefore, an ideal food plant since the entire harvest can be consumed. It has, too, the advantage that the roots can be left in the ground until needed and so can withstand insect attack and weather damage.

Cassava contains, in varying amounts, two cyanogenetic glycosides and the enzyme which hydrolyzes them to release cyanide. So-called "sweet" cassava has lower concentrations of the poisonous compounds and can be eaten after the roots are peeled and boiled or roasted. However, "bitter" cassava roots must be peeled and the flesh then grated and pounded or squeezed, the squeezing traditionally done in an elongated basket, the "tipiti," to express the cyanide-containing juice. The remaining meal is dried and made into flour or into flat breads, the original cassava. Unfortunately there are no clear-cut morphological differences which can be used to distinguish either the plants or the roots of the sweet and bitter forms.

Cassava is now widely grown in the tropics and is estimated to be the carbohydrate staple of nearly one billion people. In North America it is little-known except as tapioca, a gelatinized cassava starch, used to make bland desserts.

Sweet potato. Commonly grown with cassava was another root crop, *Ipomoea batatas,* the sweet potato. Columbus found in the Caribbean two forms, a sweet, yellow-fleshed sort called "batata," and starchier, pale-fleshed roots which he recorded as "age" or "aje." He confused both forms, as well as cassava, with true yams, the tubers of species of *Dioscorea* which he had seen in West Africa and often called them by the West African name, "niame," from which English "yam" is derived.

Sweet potatoes are undoubtedly native to the Americas, but there is good evidence for their pre-columbian spread across the Pacific. The first Europeans in Polynesia and New Zealand found sweet potatoes both as an important part of the agricultural economy and in myth and legend and it seems likely that they were carried across the Pacific some centuries before Columbus.

The roots were well-liked by Europeans who often compared the flavor to that of roasted chestnuts, and they were soon carried to other parts of the world as an easily-grown crop. They are still widely grown, both as food and as raw material for the making of alcohol, but now most of the world's sweet potatoes are produced in Asia.

Maize, beans, and squash.   When Columbus reached the West Indies *Zea mays* was the most widespread of the American carbohydrates. It was grown, in hundreds of locally adapted races, from Canada to Argentina, in a very broad range of elevations. Maize was less common in the lowland tropics where cassava and sweet potatoes were the staples, and at higher elevations in the Andes where potatoes were grown, but nearly all of the early explorers reported it. It was the staple of the great South American and Mesoamerican civilizations and was a part of religious practices throughout its range, for many maize-growing peoples had corn gods.

Its highly condensed inflorescence (cluster of flowers) is unique among grasses and for more than fifty years there has been heated controversy about the origin of maize, but there seems now to be quite general agreement that its proximate ancestor is *teosinte,* a coarse annual grass of southern Mexico. The oldest archaeological maize is dated at B.C. 5000 from Tehuacán in the Mexican state of Puebla and it is likely that maize originated in southern Mexico and neighboring Guatemala, rapidly spreading south and north.

The ear of maize is an infructescence of several hundred small one-seeded fruits, the kernels, on a common axis, the cob. This is tightly enclosed in a husk of modified leaves. Since the kernels remain attached to the cob, maize is ill-equipped to function as a wild plant, and is known only as a domesticate. Dispersal of maize requires human intervention for husking and shelling of the kernels from the cob.

Although commonly regarded as a nutritious food, maize is low in total protein and is deficient in the amino acids tryptophan and lysine, so that total dependency on a maize diet leads to nutritional deficiency diseases such as pellagra. In new world seed agriculture, maize was generally cultivated with squash and with beans of the new world genus *Phaseolus* which are high in protein and complement the suite of amino acids in maize.

The squashes associated with maize and bean agriculture belong to three or four species of the new world genus *Cucurbita.* The most widely cultivated was *Cucurbita pepo,* a variable and versatile species which provides some squashes, some pumpkins, and the ever-prolific zucchini. *Cucurbita pepo* is known from the Oaxaca Valley in south-central Mexico as archaeological material dated ca. B.C. 9700-7200; from Mesoamerica its use spread into the southwestern part of the United States and eastward to the Atlantic coast. It is likely that these species were first grown for their oil-rich seeds and that selection for thick-fleshed forms came later.

Potatoes, *Solanum tuberosum,* now the world's most important non-grain food, were first seen by Europeans in Colombia and Peru where they were grown in hundreds of varieties. Here the great advantage of the potato was its tolerance of low temperatures; it could be cultivated even at 1500 feet where little else grew. Potatoes were consumed both fresh and as the freeze-dried *chuño,* made by alternate freezing and then stamping of the thawed tubers. The dried product could be kept almost indefinitely and had lower concentrations of alkaloids than did fresh tubers.

Potatoes were slow to become accepted in Europe, in part the result of there being no tradition of use of any "root" crop as a dietary staple. Carrots and parsnips were known and used, but their use was supplemental to the staple European carbohydrates which were cereals—wheat, barley, or rye. Only in the eighteenth century did potatoes start to become widely cultivated, a result of governmental pressures and changes in population sizes and distributions.

Elizabeth Shaw

Abu-Lughod, Janet L. *Before European Hegemony: The World System A.D. 1250-1350.* New York: Oxford University Press, 1989.

Begley, Vimala & Richard D. De Puma (eds.). *Rome and India: The Ancient Sea Trade.* Madison, Wisconsin: University of Wisconsin Press, 1991.

Charlesworth, Martin P. *Trade-Routes and Commerce of the Roman Empire.* 2nd ed., revised. Chicago: Ares Publishers, 1974.

Curtin, Philip D. *Cross-Cultural Trade in World History.* Cambridge, England: Cambridge University Press, 1984.

Eisenstein, Elizabeth L. *The Printing Press as an Agent of Change.* Cambridge, England: Cambridge University Press, 1991.

French, Roger & Frank Greenaway (eds.). *Science in the Early Roman Empire: Pliny the Elder, His Sources and Influence.* London: Croom Helm, 1986.

Heiser, Charles B. *Of Plants and People.* Norman: University of Oklahoma Press, 1985.

Morton, A.G. *History of Botanical Science.* New York: Academic Press, 1981.

Parry, John H. *The Age of Reconnaissance.* Berkeley: University of California Press, 1981.

Pullapilly, Cyriac K. & Edwin J. Van Kley (eds.). *Asia and the West: Encounters and Exchanges from the Age of Explorations.* Notre Dame, Indiana: Cross Cultural Publications, 1986.

Sauer, Carl O. *The Early Spanish Main.* With a new foreword by Anthony Pagden. Berkeley: University of California Press, 1992.

Simpson, Beryl B. *Economic Botany: Plants in our World.* New York: McGraw-Hill, 1986.

The Exhibition

1   CHRISTOPHER COLUMBUS (1451–1506)

Epistola de insulis noviter repertis.

Paris: Guy Marchant, 1493.
The Houghton Library. Gift of the curators of the Bodleian Library, Oxford University, in September, 1936, on the occasion of the three-hundredth anniversary of the foundation of Harvard College.

The first news of the new world came in the "Columbus letter." It is dated on board the *Niña*, off the "Canary Islands"—in fact, the Azores—on the 15th February, 1493. Columbus' own journal of the voyage is lost and is known only from its use by Ferdinand Columbus in his biography of his father (Venice, 1571) and by the abstract made by Bartolomé de las Casas which was not printed until 1825.

This first work on the new world was printed at Barcelona in the spring of 1493 and at Rome, in Latin, a month or so later. By the end of the year, the letter had been printed three times more at Rome, once at Antwerp, once at Basel, three times at Paris, and twice at Florence. In 1495 it was twice printed at Florence; in 1497 a German translation was printed at Strasbourg and there was a second printing in Spanish at Valladolid.

We know from the abstract that Columbus kept a surprisingly detailed journal which contains much information about plants of the newly-found land. We find the first mention of new world cotton (13 October), of maize (16 October), of sweet potatoes and new world beans (4 November), and of tobacco (6 November). Columbus also recorded products of more immediate interest, cinnamon (4 November), mastic (5 November), and aloes (14 November). None of these are native to the new world, but Columbus saw plants which he thought were the things he so hoped to find.

In the letter he summarizes his discoveries, emphasizing the docility and gentleness of the people, greatly overemphasizing the little gold he received, and describing a beautiful and fertile landscape. "La Spañola [Hispaniola] is marvelous…the lands are so beautiful and rich for planting and sowing…there are many spices and great mines of gold and other metals." Columbus wished to present to Ferdinand and Isabella proof that he had reached the Indies—and what better proof could there be than a land rich with spices?

He offers to the King and Queen as much spice and cotton, and gum mastic and aloe as they would wish to have shipped. "And I believe that I have found rhubarb and cinnamon, and I shall find a thousand other things of value." Cotton he did find, but the "spice" was probably *Capsicum*, chili pepper, not the pepper from India. The mastic, the aloes, the rhubarb, the cinnamon—all of these were dreams. They are from the other world, not this new one, which would offer products of value far greater than all these combined.

Hindman, Sandra. "The Career of Guy Marchant (1483–1504): High Culture and Low Culture in Paris." In S. Hindman (ed.), *Printing the Written Word*. Ithaca and London: Cornell University Press, 1991.

Morison, Samuel E. *Journals and Other Documents on the Life and Voyages of Christopher Columbus*. New York: The Heritage Press, 1963.

De orbe novo.

Alcalá de Henares: M. de Eguia, 1530.
The Houghton Library.

Peter Martyr, as his name is anglicized, may have been the first man in Spain to realize how important was the discovery made by Columbus. Born near Milan, Martyr spent his early years there and in Rome where he attained some fame as a scholar. In Rome he met the Spanish ambassador, and when the ambassador returned in 1487 to Spain, Martyr was persuaded to come with him. Apart from a trip to Egypt in 1501–02, Martyr spent the rest of his life there.

In Spain, Peter Martyr took minor orders and soon became a chaplain in the royal household, a position which assured him both a comfortable life and proximity to the most interesting events of that eventful time. In November, 1493, he began writing to friends in Rome that series of letters about the new world which is known as the "oceanic" decades.

Martyr has been called a gossipy cleric, but the "gossipy" is unfair. He was a most intelligent and inquisitive man, interested in every piece of news about the new world which reached the royal court. The best of these are included in his letters which provide a marvelous description of the early years of the new world. Martyr knew many of the early explorers and conquistadors; he knew Columbus, he knew Vicente Pinzón, and Fernández de Enciso. He questioned them about everything they could re-call and he passes along in the letters what his informants had to say about the events in which they had been participants, and about the land, its people, and its plants and animals. The letters read as if he had been there, although he never was, and they give a vivid picture of the first thirty years of America.

The letters were widely read and, after books compiled from them had been printed without his authorization, Martyr saw that they were printed under his own direction. Some of the decades were printed earlier, but *De orbe novo* is the first complete printing of all eight of them. The first three de-cades cover the years from 1492 through 1516. Then for a few years Martyr did no writing, but the news of Magellan's circumnavigation and of Cortés' conquest of Mexico led him to resume the letters.

The first two letters of the first decade are addressed to Cardinal Ascanio Sforza in Rome. The first letter was written in November, 1493, a few weeks after Columbus had left Cadiz on his second voyage. Here Martyr tells the Cardinal of the foods used in the Indies—strange root crops, such as sweet pota-toes, and another called "yuca," which is manihot, along with a description of its preparation and its use as bread. He mentions maize which he compares to millet. In the third letter, dated May, 1500, to Cardinal Ludovico d'Aragon, Martyr reports upon the fertility of Hispaniola; all of the kinds of seeds brought from Europe germinated and grew amazingly quickly.

The plant which most interested Peter Martyr was cacao, discovered in Mexico, where the seeds were used both as currency and to make the drink, chocolate. He remarks that it is a blessed money, for it not only provides a delightful drink, but it prevents its possessors from yielding to avarice, since it cannot be kept for any length of time. Martyr was interested, too, by the Mayan ball game which he calls tennis. He was intrigued by the bouncing balls, which he says are made of the sap from a vine which climbs as do hops. The vine may have been a species of *Castilloa*, a latex-producing genus of the fig and mulberry family. This is the first mention in European literature of rubber, although it is not the familiar modern rubber made from the latex of *Hevea brasiliensis*.

3   FRACANZIO DA MONTALBODDO (fl. 1507)

Paesi novamente retrovati & Novo Mondo da Alberico Vespatio.

Milan: G.A. Scinzenzeler, 1508.
The Houghton Library.

As word spread in Europe of the discoveries in America and in the east, there came demands for more information about the riches and wonderful things found in these two new worlds. The first collection of narratives of travel, *Libretto di tutta la navigazione del Re di Spagna,* was printed at Venice in 1504. The *Libretto* usually is attributed to Angel Trevisan who was secretary to the Venetian ambassador to Spain, and so had easy access to his source material. Included in the *Libretto* are accounts of the first three voyages of Columbus and of the first voyage of Vicente Yañez Pinzón who had been captain of the *Niña* in 1492. It is, in fact, an unauthorized—pirated—printing of the early letters of Peter Martyr.

The *Libretto* was printed once and is now known by two copies, but the narratives there were used three years later in the *Paesi,* first printed at Vicenza in 1507. Compiled, it is thought, by Alessandro Zorzi of Vicenza, and brought to press by Montalboddo, the *Paesi* contains accounts of voyages to west Africa, a narrative of Cabral's voyage of 1500–01, and letters describing da Gama's expeditions to India.

The Americana are the Columbian voyages, Pinzón's voyage, the voyage of Gaspar Corte Real to Newfoundland, and *Mundus Novus,* first printed in 1503, which tells the story of Vespucci's voyage (1501–02) to Brazil. The *Paesi* was soon translated into Latin, French, and German and was being reprinted, with added material, as late as the 1530's. In the first editions Europe found, suddenly laid out before it, an entire new world from Newfoundland in 50° north, to Brazil and the equator, all of it unknown to Europeans only fifteen years before.

The *Paesi* is a handsome little book and as one pages through it, nearly 500 years after its printing, there lingers still a sense of the excitement felt by its first readers. As Boies Penrose, historian of Renaissance travel and exploration, said of it, the *Paesi* was one of the most influential books ever printed.

The first three books deal with navigations in the old world and at the end of the third book are lists of the values of commodities brought from Calicut on the Malabar coast of India—brazilwood, cassiafistula (a commonly used purgative), cloves, cinnamon, rhubarb and others. As the reader turns the page to book four to find "Incomenza la navigatione del Re di Casteglia de la Isole & Paese novamente retrovate" these valuable products are clear in his mind. Perhaps they occur in these newly found lands of the King of Spain—if not these, then others the same or even better.

Pinzón's voyage landed him in January, 1500, somewhere on the "bulge" of Brazil. For five months he headed northwestward, even sailing a short distance into the mouth of the Amazon, into the Gulf of Paria and returned to Spain by way of Hispaniola. It had not been an easy sail and Pinzón lost nearly half of his men, but he returned with brazilwood and reported finding both brazilwood and cassiafistula in quantity, along with pearls and herbs, and plants, and trees, and animals strange and diverse. Columbus had found these things, but very little about his voyages had been printed, and that little had nothing like the distribution that the *Paesi* had. To most of its readers this was new and exciting information.

Penrose, Boies. *Travel and Discovery in the Renaissance, 1420–1620.* Cambridge, Massachusetts: Harvard University Press, 1952.

4   MARTÍN FERNÁNDEZ DE ENCISO (fl. 1509–30)

Suma de geografía.

Seville: J. Cronberger, 1519.
The Houghton Library.

Enciso's *Suma* was prepared as the authoritative work on world geography. In 1513, Nuñez de Balboa had looked down upon the Pacific Ocean and it was clear that Balboa's ocean was the same as that which Portugal was investigating from the other side. It was time for a comprehensive work which would lay out the world as it was known after more than fifteen years of exploration both east and west.

Enciso, a lawyer, spent several years in the new world. Sometime before 1509 he had gone to Hispaniola where, as a successful lawyer and, later, a judge, he became financially involved with Alonso de Hojeda who had been granted a concession to hold what is now the northern coast of Colombia. After Hojeda's attempt to establish a settlement on the Gulf of Urabá failed, he fled to Hispaniola. Enciso was sent as his successor, but the scholarly judge was ill-equipped to be the leader of a frontier camp with characters such as Francisco Pizarro and Balboa and he was forced to return to Spain. Enciso made another short visit to the Caribbean, but by 1517 he was permanently back in Spain.

The *Suma* is a practical work meant to serve as a guide for navigators and Enciso works his way along the coasts of the known world, giving distances and latitudes. Longitudes would not be measured with any accuracy until late in the eighteenth century. In the colophon Enciso lists his sources, among them Ptolemy, Pliny, Strabo, and the Bible, all useful for the old world, and "the experience of our times which is the mother of all things." This Enciso could provide for the new world. He devotes nearly nineteen pages to America; fourteen lines deal with North America which, so far as Enciso was concerned, consisted of Labrador and codfish.

In this utilitarian work Enciso presents useful information on poisonous plants, about plants of commercial value, such as brazilwood and cassiafistula, and on food plants. The poisonous plant Enciso mentions is the mancineel, *Hippomane mancinella,* known in tropical America for its toxic latex, which can cause temporary or even permanent blindness. Enciso explains that anyone who even sleeps in the shade of a mancineel tree can awake with his eyes swollen shut. This is true, but then he repeats, or perhaps starts, the false tale that the fruits when eaten turn to worms in the body.

In the *Suma,* Enciso gives the first explanation of the differences in foods between the islands and the mainland. He notes that, from Cartagena westward, bread, as well as a fermented drink, was made from maize. There were also roots used, such as were eaten in Hispaniola, but these on the main land could be eaten cooked or even raw; the roots were *batatas* (sweet potatoes), *ages* (white-fleshed, starchy sweet potatoes) and "sweet" cassava. Enciso offers the note of caution that in the islands the cassava ("bitter" cassava) was poisonous unless the roots were grated and the juice squeezed out. The treated meal was then made into flat breads.

The geographical distinction was invaluable information for explorers who could not know that cassava contains two cyanogenic glycosides and the enzyme which hydrolyzes them. When the flesh is cut or even bruised, prussic acid (hydrogen cyanide) is released. Usually this is removed by grating the roots and expressing the juice. Cooking then drives off the last traces of cyanide. "Sweet" cassava contains a smaller quantity of the glycosides, which are concentrated in the outer layers of the roots and so can be removed by peeling. Unfortunately there are no morphological differences between the sweet and bitter forms and there must have been a great deal of trial and disastrous error.

Sauer, C.O. *The Early Spanish Main*. Fourth printing, with a new foreword by Anthony Pagden. Berkeley: University of California Press, 1992. An excellent account of the Spanish new world from 1492 to 1519.

5   Giovanni da Verrazzano (1480?–1528?)

"Relatione di Giovanni da Verrazzano Fiorentino della terra per lui scoperta…" in G.B. Ramusio, *Terzo volume delle navigationi et viaggi.*

Venice: Giunti, 1556.
The Houghton Library. Gift of John B. Stetson, Junior, of the Class of 1906.

By 1520 it was clear to those watching Spanish activities in the new world that no rich mines of gold or silver had been found, and that, if money were to be made there, it would be from products of the land, not from minerals. Of course, Cortés had not yet taken Mexico and Potosí had not yet been discovered. It was clear, too, that America was not the Far East and, to some, it was only an obstacle in the way to a westward approach to the orient.

In 1523, a group of Florentine silk merchants living in Lyon pledged funds to finance an expedition in search of a passage through the new lands to Cathay. This kind of commercial support of an exploration was not new, and the venturers would share in any profits in proportion to their subventions. This expedition was led by Giovanni da Verrazzano, of Florentine origin, who was probably closely related to some of the backers.

Our knowledge of the voyage comes from a letter addressed to Francis I who provided the ship. The original letter is lost but a copy was available to Giovanni Battista Ramusio who included it in the third volume of his magnificent collection of narratives of travel.

Verrazzano sailed with a single ship from Madeira in January, 1524 and sometime in March made landfall at or near Cape Fear, North Carolina. He describes the country, "with many beautiful fields and plains full of great forests…they are adorned and clothed with palms, laurel, cyprus and other varieties of trees unknown in our Europe….We think that they belong to the Orient by virtue of the surroundings, and that they are not without some kind of narcotic or aromatic liquor."

From Cape Fear, Verrazzano sailed up the coast, visiting New York harbor, which he named Angolême [sic], where he saw the Lower Bay and the narrows which still carry his name. Verrazzano considered New York to be "a very agreeable place between two small but prominent hills." From New York they continued along the outer coast of Long Island and into Narragansett Bay (which Verrazzano called "Refugio"), where the crew spent a delightful fifteen days at what is now Newport, Rhode Island. They then rounded the dangerous shoals off Nantucket Island and ran up the coasts of Massachusetts and Maine. At about 50° north, Verrazzano, running low on supplies, turned back to France, and sent to Francis I the letter dated 8 July, 1524.

Verrazzano's descriptions of the people are full of detail; at Cape Fear he noted that they were wearing clothing made of Spanish moss *(Tillandsia usneoides):* "certain grasses that hang from the branches of the trees." As for food plants, several times the people are said to live on pulses, "legumi," but there is no mention of maize.

Unfortunately for his backers in Lyon, Verrazzano brought back neither a passage through the continent nor gold and spices. However he did fill in on the map the coast between Florida and Newfoundland, and Boies Penrose called his account "the most accurate and the most valuable of all the early coastal voyages that has come down to us."

Parks, George B. "The Contents and Sources of Ramusio's *Navigationi.*" *Bulletin of the New York Public Library* 42: 279–313. 1955.

Penrose, Boies. *Travel and Discovery in the Renaissance, 1420–1620.* Cambridge, Massachusetts: Harvard University Press, 1952.

Wroth, Lawrence C. *The Voyages of Giovanni da Verrazzano, 1524–1538.* New Haven: Yale University Press for the Pierpont Morgan Library, 1970.

Early image of a hammock. From a reprint of G. F. de Oviedo y Valdés, *Historia general y natural de las Indias* in G.B. Ramusio, *Terzo volume delle navigationi et viaggi,* Venice, 1556. (See catalogue item number 6.)

6   GONZALO FERNÁNDES DE OVIEDO Y VALDÉS (1478–1557)

De la natural hystoria de las Indias.

Tolédo: R. de Petras, 1526.
The Houghton Library. From the bequest of the Hon. William Prescott.

By 1530 anyone who wished to know something of the new world would have been able to get a very good picture from Peter Martyr's *De orbe novo* (1530), with its excellent, though second-hand reports, and from Fernández de Oviedo's *De la natural hystoria de las indias,* written by a man who had lived there and who loved the land.

Oviedo went in 1514 to Panama as supervisor of gold-smelting for the Emperor Charles V. He spent the next forty years in the new world, going several times back and forth to Spain, and returned finally in 1556 to Valladolid where he died. Oviedo is best known for the great *Historia general y natural de las Indias* (Seville, 1535; Valladolid, 1557). However, in 1524 Oviedo was in Spain waiting for an audience with the Emperor. At the request of Charles, Oviedo wrote this shorter *Natural History of the Indies.* Oviedo was an intelligent and interested observer and his descriptions are factual, with little overlay of the marvelous and wonderful.

The larger work contains much political history, but *De la natural hystoria* is just that, with information about the plants and animals of the land and much information on the people, their customs, and their foods. The emphasis is on Hispaniola and Panama, the areas which Oviedo knew best.

Oviedo gave his readers the first eyewitness accounts of many new world plants. His descriptions are good, for he catches the characters which are quick identifiers, as "in Tierra Firme there are trees called pear trees…this tree bears pears which weigh about a pound…the juice and the flesh taste much like butter…these pears are very good with cheese." There is no doubt that the "pear tree" is *Persea americana,* the avocado. He describes, too, *Guaiacum officinale,* the guayacán or palo santo of Hispaniola, tells how it is used in treating syphilis, and plainly says that the disease was brought from the new world on the first two voyages of Columbus, and that the disease was new to Europe.

He gives an excellent description of pineapples which are, Oviedo remarks, produced on plants like aloes. Of all the new fruits which Europeans met in America, the pineapple was the favorite. Everyone, and Oviedo is not an exception, describes, often with near rapture, its flavor and odor. Oviedo reports that it is one of the best fruits in the world and that an entire house can be sweetly scented by one or two fruit.

Oviedo provides the first detailed descriptions of maize and of cassava which provided the new world's breads and explains how "bitter" cassava had to be prepared for bread-making by grating and then squeezing out the cyanide-containing juice. Cassava bread was highly regarded, for while maize bread became moldy in a few days, cassava bread could be kept, Oviedo says, for more than a year, and was frequently used as ship's supplies.

In reading all that Oviedo says about the plants and animals, one notes running throughout a disturbing thread—the rampant growth and spread in the new world of species which had been brought from Spain. Only thirty years after the discovery there were so many bananas, which multiplied so rapidly, that Oviedo says that one hardly could believe it without having seen it. Of cattle he reported that some "cattle kings" had herds of 3,000–4,000 head, and herds of 500 were quite common. The same had happened with swine; feral pigs were everywhere. As Oviedo noted of European plants, "they grow and multiply, in spite of being neglected and not cared for."

Fernández de Oviedo, G. *Natural History of the West Indies.* Translated and edited by S.A. Stoudemire. Chapel Hill, North Carolina: The University of North Carolina Press, 1959.

Gerbi, Antonello. *La natura delle Indie Nove: Da Christoforo Colombo a Gonzalo Fernándes de Oviedo.* Milan: Riccardo Ricciardi Editore, 1975. Translated by Jeremy Moyle as *Nature in the New World.* Pittsburgh, Pennsylvania: University of Pittsburgh Press, 1985. A unique and invaluable work on its subject.

7   ALVAR NUÑEZ CABEZA DA VACA (ca. 1490–1550)

La relación y commentarios del governador Alvar Nuñez Cabeça de Vaca.

Valladolid: Francisco Fernández, 1555.
The Houghton Library. From the bequest of the Hon. William Prescott.

In February, 1527, Pánfilo Narváez sailed from Cuba with five ships and 400 men. These few facts introduce the most ill-fated of sixteenth-century explorations, for Narváez had already lost 200 men in the eight months since leaving Spain with the goal of conquering and colonizing "Florida," a term which included the peninsula and much of the Gulf coast.

On the first of May, Narváez took possession of Florida in the name of Charles V. He found not a rich land and a great city, as had Cortés, but thick forest and swamps, and there began a terrible series of disasters, brought by bad luck and, in part, by inept management. By December, 1528, Narváez was dead and fewer than 100 men remained, shipwrecked on "Malhado" [Bad Luck] Island, perhaps Galveston Island off the Texas coast.

By the spring of 1529 only fifteen were alive. In the end, out of Narváez' original 600 men, four were left—Nuñez Cabeza de Vaca, Andrés Dorantes, Alonso del Castillo, and Estévan, Dorantes' Moroccan slave. All were barefoot and naked for they had lost everything in the final shipwreck. For four years they were on the Texas coast, "earning" a chancy existence, not a living, as faith-healers, breathing on the ill and saying for them (and for themselves) a few prayers.

Finally the four left to walk across Texas towards Mexico, healing as they went. The country they crossed was inhabited by people who grew few crops, but lived by gathering and some hunting. Cabeza de Vaca and his companions could live no better than their hosts and usually were much worse off, for they were treated as near slaves. Often their food was only a little parched maize, or, the fruit of species of cacti (prickly pears), or even only the roasted stem segments of cacti *(Opuntia)* with a little protein from whatever animals the Indians could kill. Only when they were in trans-Pecos Texas did they find people who grew maize, beans, and squash.

In the spring of 1536 they met at last a party of Spanish raiders somewhere in the Mexican state of Sinaloa. In August of the next year Nuñez was back in Spain and his story was first printed at Zamora in 1542. The book here on exhibit includes, as well, *Los Commentarios* of Nuñez' years (1541–42) on the Río de la Plata.

In Nuñez' narrative readers found a new reality of America—not an Edenic garden of gentle people and fragrant breezes, but a harsh landscape in which Europeans were not immediate conquerors. It demonstrated, too, how enormously broad was the continent, a realization accepted by Spain long before the English came to understand it.

8   GIROLAMO BENZONI (1519–72?)

La historia del Mondo Nuovo.

Venice: F. Rampazetto, 1565.
The Houghton Library.

Girolamo Benzoni of Milan, a silversmith who wished to see the world, went in 1541 to America. During his fifteen years in the new world Benzoni lived and travelled in Cuba, central America, and in

Peru. Having acquired some money Benzoni decided at last to return to Europe and in 1556 was back in Milan. His book stands apart from other sixteenth-century narratives of travel in the new world in Benzoni's clearly and strongly expressed feelings about excesses against the Indians.

In 1578 was published a Latin translation and in 1579 a French translation of Benzoni's book, both by the French Huguenot pastor, Urbain Chauveton (ca.1540–ca.1614). Together with the *Brevissima rel-*

Making of maize bread. From Benzoni, *La historia del Mondo Nuovo.* Venice, 1565. (Catalogue item no. 8.)

Cacao tree in fruit; in background seeds being dried. From Benzoni, *La historia del Mondo Nuovo.* Venice, 1565. (Catalogue item no. 8.)

Caciques in tree house under Spanish attack. From Benzoni, *La historia del Mondo Nuovo.* Venice, 1565. (Catalogue item no. 8.)

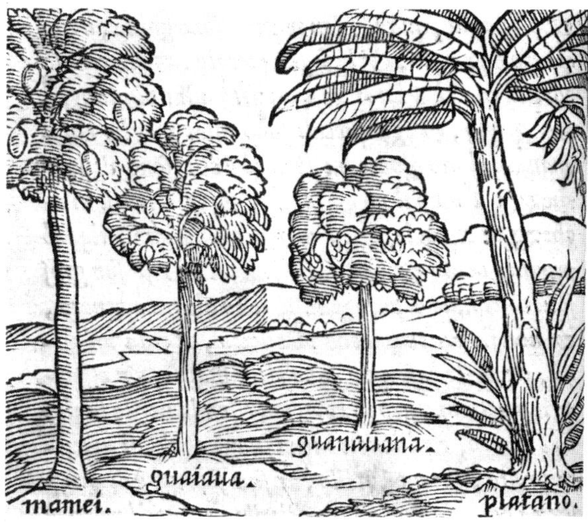

Fruits of the new world: mammey, guava, soursop, and banana (but the latter was native to the old world). From Benzoni, *La historia del Mondo Nuovo.* Venice, 1565. (Catalogue item no. 8.)

acion de la destruycion de las Indias of Bartolomé de las Casas, these translations, with Chauveton's annotations, played a large part in spreading the "black legend" of Spanish cruelty to native peoples. Chauveton's Latin translation was given further currency by its inclusion in Theodore de Bry's "Grands voyages"—in three installments, part 4 (1594), part 5 (1595), and part 6 (1596).

Benzoni's book is rich in information about events, about the native people, and about the plants and animals which filled the landscape. Many of his discussions of the commonly used plants are excellent and amazingly detailed—but it is made clear that Benzoni likes very few of the new world's products. Why, we wonder, did he spend fifteen years there? The only American plant which receives unreserved praise is the pineapple, "the best fruit in the whole world." Everything else is more or less strongly criticized.

Benzoni carefully explains how cassava bread is made and notes that all ships returning to Spain load it for their stores, but he calls it "a wretched article of food." The planting of maize is described— make a small hole, put in three or four kernels, and cover—and its grinding into flour to make tortillas. But, Benzoni says, maize is good neither hot nor cold. He explains how cacao is grown, under the shade of larger trees, and how the drink is prepared. But "it is better suited for pigs than for men."

Benzoni's strongest words are reserved for tobacco. He describes the making of cigars by wrapping dried leaves of tobacco in a maize leaf, and says that the Indians use so much that they fall down stupefied. Then comes his blast. In Chauveton's Latin, "Quam pestiferum, obsecro, et noxium hoc Tartareum venenum est." In the words of W.H. Smyth, who translated Benzoni for the Hakluyt Society edition, "See what a pestiferous and wicked poison from the devil this must be!" The effects of tobacco smoking have often been described, but this seems to be the first anti-smoking statement.

Benzoni, Girolamo. *Historia del nuevo mondo.* Introduction and notes by Manuel Carrera Diaz. Madrid: Alianza, 1989.

Keen, Benjamin. "The Vision of America in the Writings of Urbain Chauveton." In *First Images of America,* 2 vols., edited by Fredi Chiappelli, Volume I. Berkeley: University of California Press, 1976.

9   Pédro de Cieza de Léon (1518–60)

La chronica del Peru.

Antwerp: M. Nacio, 1554.
The Houghton Library.

Cieza de Léon, born at Seville, went at the age of thirteen or fourteen to the new world, perhaps with the expedition of Pédro de Heredia in 1532. Here he became a soldier and travelled widely in Colombia and Peru, both in the Andes and along the coast. Cieza tells us that in 1541, while at Cartago in the Cauca valley of Colombia, he decided to keep a journal and to record those things he saw.

For a man who, hardly more than a child, had set off to the new world and who spent his teenage years in the Peru of the 1530's, Cieza was an intelligent and sympathetic observer who was interested in and described anything from funeral rites to fruit trees. After seventeen years in America, Cieza returned in 1550 to Spain. His chronicle of Peru was intended to be in four sections, but only the first part dealing with geography and natural resources was printed, first at Seville in 1553. It was reprinted the next year in Antwerp and the year following at Rome in Italian.

As Cieza describes his travels, there is a steady flow of information about the plants used by the native people. He gives us first mention of the Andean potato which Cieza describes as a kind of earth nut, growing underground like a truffle, which when boiled is as tender as chestnuts; he also describes how *chuño,* the freeze-dried potato food of the Andes, is made. Potatoes and *quinoa* were the principal carbohydrate food of people who lived above the valleys.

Maize was also grown at lower elevations and Cieza notes the fact that impressed so many of the early writers—the enormous yield of the crop. Cieza marvels that a fanega (about one and a half bushels) of maize will return a hundredfold or more.

At first reading, a hundredfold might seem an exaggeratedly high yield. But in examining contemporary images of maize, one sees plants which bear two to four ears, each with 250 to 300 kernels. If one assumes an average yield of three ears per plant, with two of the ears kept for food, each plant then yields 250 or more kernels for sowing. If only half of these develop into mature fruiting plants, each with three ears, we already have more than a hundredfold return, jumping in one generation from 3 to 375 ears, and in many areas maize was sown twice a year.

When Cieza did his notetaking, Europeans had been in Peru for fewer than twenty years, and one is surprised to find so many crops of the old world already being grown there. Around Quito there was much wheat and barley, and at Pasto barley was commonly grown. In Colombia, at Cali, settled only in 1536, there were being grown oranges, limes, lemons, pomegranates, bananas and sugar cane, pretty much the array of crops one might have seen in contemporary Andalusia and Granada.

10   Garcilaso de la Vega (1539–1616)

Commentarios reales que tratan del origen de los Incas.

Lisbon: P. Crasbeeck, 1609.
The Houghton Library.

Garcilaso de la Vega, son of Sebastián de la Vega Vargas and Nusta Chimpu Ocllo, an Incan princess, was born in Cuzco. At the age of twenty-one, the young man left Peru for Spain where he spent the rest of his life. Baptized Gómez Suárez de Figueroa, it was only after his arrival in Spain that he began to use the name Garcilaso de la Vega after his distant relative, the poet of that name, and it was to distinguish the two that "El Inca" became his byname.

Garcilaso's *Commentarios Reales* is a remarkable work, written by a man who had a very personal experience of Peru during the years just after the conquest. Much of the history of the first years after the conquest is taken, as Garcilaso acknowledges, from writers such as Zarate, López de Gómara, and Cieza de Léon, but he gives, in addition, a great deal of information on the life and customs of the people taken from his first-hand knowledge, for he was raised among his mother's people.

He mentions, for example, that maize was used not only as a substantial foodstuff, but, by the Spanish at least, as a medicine. They found maize "good" for bladder and kidney stones: Garcilaso notes that few Indians had such problems and that the Spanish believed this to be a benefit from drinking a brew of maize—whether fermented or not, de la Vega does not say.

In Book Eight, de la Vega has several chapters on food plants, some of which, he remarks, grow above ground and others below. Above ground the most important was, of course, maize, of two kinds, one hard, the other "soft and very tasty." The soft one was eaten boiled or roasted—the hard sort was ground into flour and made into bread. Also mentioned are several kinds of beans, and quinoa, used both as a pot-herb and as a grain.

Below ground grow potatoes, which de la Vega says are used in those regions which do not grow corn, as well as sweet potatoes and peanuts. Peanuts, he notes, cause headaches if eaten raw so they must be toasted; they were used, too, as a source of oil.

Highly regarded were chilies, or *uchu*. Garcilaso says that they take first place among "fruits" and notes that the "inhabitants of my own country are so attached to uchu that they will eat nothing without it." Chilies were so esteemed that, when observing a fast, giving them up made the sacrifice the more meaningful. Indeed, when young Incas, as candidates for manhood, went through the initiation ceremonies, they fasted for six days, consuming only maize and water, but no chilies.

Concerning coca, de la Vega quotes Blas Valera. Valera lists the value of coca as a medicine and then—"It has another great value which is that the income of the bishop, canons, and other priests of the cathedral church of Cuzco is derived from the tithe on the coca leaf, and many Spaniards have grown rich, and still do, on the traffic in this herb." Garcilaso himself later explains that trade in carrying coca and Indian textiles from Cuzco to Potosí was a desirable employment for Spaniards, however noble, since it did not involve selling goods in shops, and so lacked the stigma associated with trade.

11   José de Acosta (1540–1600)

Historia natural y moral de las Indias.

Seville: J. de Léon, 1590.
The Houghton Library. From the fund for a Professorship of Latin-American History and Economics.

Acosta, born at Medina del Campo, joined the Society of Jesus as a young man, and in 1570 was sent to Peru where he stayed until 1586. Returning to Spain with many years' accumulation of manuscripts, he quickly arranged their publication. Most of his works dealt with theology, but the greatest and best known of all is *Historia natural y moral de las Indias* which is an extensive and valuable description of the natural history of Peru and to a lesser extent of Mexico, where Acosta spent a year on his way back to Spain.

In Peru the Jesuits had a center at Juli on the western shore of Lake Titicaca where Acosta may have been based, but he travelled widely and clearly was a keen and interested observer of his world. He describes the principal crops of Peru, maize where it could be grown, and at the higher elevations, potatoes which, Acosta says, are grown in cold, high, dry areas and often freeze-dried into *chuño*—as they still are. On the subject of maize, Acosta quotes Francisco de Tolédo, viceroy for several years, to the effect that Peru had two valuable products, maize and cattle, the "cattle" being llamas, valued for their wool.

As did every other chronicler of Peru, Acosta discusses coca which was used, especially in the mines of Potosí, in remarkable quantities. He explains that it was grown in the valleys and taken by llama trains to the mines where 90,000 to 95,000 thousand baskets of coca were used every year; in 1583, 100,000 baskets were consumed by the miners. Acosta says that there was much discussion about tearing out the coca plantations, "but in the end they remained."

Acosta discusses or, at least, mentions most of the plants used in Peru as foodstuffs or as medicinals, and even the ornamentals. He remarks that the Indians loved flowers just for their beauty, and mentions *Datura,* sunflowers, passion flowers, and another, beautiful to look at, but unscented. This one would remain unknown to us were it not that Acosta says that the plant tastes just like cress—it is our garden nasturtium, a species of *Tropaeolum.*

For all that Acosta says about the plants of Peru, the most impressive facts are these. He returned to Spain with the bullion fleet of 1587, and several times mentions the cargo carried. In that one sailing were carried 11,000,000 pieces of silver—which one would expect. But there were also 22,053 hundredweight (quintals) of ginger, 50 quintals of sarsaparilla, 48 quintals of cassiafistula, 350 quintals of guaiacum, and 1309 hundredweight of brazilwood. Added to this were nearly 100,000 hides picked up at Santo Domingo—and this was in a single year.

Ginger and cassiafistula are products of old world species, early brought to Hispaniola and, by Acosta's time, widely grown, but sarsaparilla, guaiacum and brazilwood are native species. We might note that the native species were not cultivated, but were wild-collected, and destructively so; if sarsaparilla, of which the roots were used, it was dug up; if guaiacum or brazilwood, the trees were cut down.

12 HANS STADEN (1525?–70?)

Warachtighe historie ende beschriivinghe eens lants in American gelegen.

Antwerp: J. Roelants, 1563.
The Houghton Library. From the bequest of the Hon. William Prescott.

Two of the most interesting narratives of new world travel were written by "simple soldiers," that of Cieza de Léon on Peru and that of Hans Staden on his adventures in Brazil. That Staden's story was told at all is the more remarkable for he was held prisoner for months by Tupinambá Indians who threatened on a near-daily basis that he would be roasted and eaten.

Eager to see the world, Staden went as a young man to Lisbon and shipped out as a gunner on a ship carrying convicts to Brazil. On this first voyage (1547) Staden was gone from Portugal for sixteen months. Anxious to see the Spanish lands in the new world, Staden sailed again on a ship bound for the Río de la Plata. But after a tedious and long voyage Staden found himself again in Brazil, and eventually in command of a Portuguese fort at Santos.

Here he was captured by a raiding party of Tupinambás who regarded the Portuguese as their archenemies because of their slaving activities. Although Staden tried to explain that he was German—and had a red beard rather than a Portuguese black one to prove it—he was held for more than nine months, escaping at last to a French ship. In 1555 he was home in Germany and must have turned at once to preparing his book.

The book was first printed at Marburg in 1557 and was one of the earliest accounts of the new world in German. With its detailed accounts of Staden's narrow escape from being eaten, it was an instant success, and was reprinted and translated into Latin, French, Dutch, and Flemish. Staden was an intelligent man who wrote a good, straightforward narrative of his adventures.

The book is in two parts; the first is Staden's narrative of his two voyages, most of it devoted to the second trip during which he was held prisoner. In the second book Staden describes, as he saw it, the country around Santos and the way in which the Tupinambá lived. They were hunters and slash-and-burn agriculturalists whose chief carbohydrate food was manioc.

Staden gives an excellent description of manioc preparation—how the roots were ground or grated and then placed into an elongated cylindrical basket which Staden calls "tippiti," the name still used today. This was used to squeeze out the cyanide-containing juice, and the dried meal, cassava, was made into cakes. The Indians collected manioc and made cakes on a daily basis and this was the pattern followed by Europeans as well. On Staden's first voyage he spent time at a small settlement near Olinda and found that the Tupis knew very well that Europeans could quickly be starved out by keeping them from the manioc fields.

Europeans, especially the French, were early fascinated by the handsome, feather-adorned, innocent cannibals who lived in a near-Paradise. Staden's book and that of André Thevet on the French settlement at Rio de Janeiro were printed in 1557 and both were enormously popular for they offered close-up views of an exotic and dangerous land.

Hemming, John. *Red Gold: The Conquest of the Brazilian Indians.* Cambridge, Massachusetts: Harvard University Press, 1978.

Staden, Hans. *The True History of his Captivity.* Translated and edited by Malcolm Letts. London: George Routledge & Sons, Ltd., 1928.

13   ANDRÉ THEVET (1517?–92)

Les singularitez de la France Antarctique.

Antwerp: C. Plantin, 1558.
The Houghton Library. Gift of Samuel A. Eliot.

Thevet's book is founded upon a strange interlude in the history of Brazil. The coast was discovered, perhaps by Vespucci in 1499, certainly by Pinzón and Cabral in their voyages of 1500, but the first settlements were not made until 1531–32 at Bahia in the north and at Santos in the south. During these thirty years, ships of several countries, especially France, called in and cut loads of brazilwood, used to make a valuable dye.

In 1555 Nicolas Durand, Chevalier de Villegagnon, a professed Protestant, persuaded Henri II of France that a French colony could be established as a possible refuge for Huguenots should the uneasy and tense truce between Catholics and Calvinists break apart.

André Thevet, a Franciscan who had experience of travel, for he had spent four years in the Levant, was appointed chaplain to this Protestant-led expedition. Soon after Villegagnon's party landed at Guanabara Bay (Rio de Janeiro) in November, 1556, Thevet fell ill and returned to France sometime in January, so that his personal experience of Brazil was limited to his few months at Guanabara Bay. He was quickly in print (1557) with *Les singularitez de la France Antarctique,* "antarctique" used to refer to regions south of the Tropic of Capricorn.

Two books on Brazil were printed in 1557, that by Thevet and Hans Staden's *Warhafftige Historia* (see previous entry). Both were very popular, especially that of Staden with his details, usually gory, about cannibalism. *Les singularitez* was reprinted in 1558 at Paris and at Antwerp, and a few years later Italian and English translations were published at Venice and in London.

Indians making fire and smoking. From A. Thevet, *Les singularitez de la France Antarctique.* Antwerp, 1558. (Catalogue item number 13.)

Tupinambá Indians wearing and making ankle bracelets from the fruit of *Thevetia ahouai;* the man to the left holds a *maraca* or sacred rattle.

Gathering fruit of *Crescentia cubete*, the calabash tree.

Gathering "cashew apples" (enlarged fruit stalks) for winemaking.

"*Hetich*," the sweet potato.

Felling of brazilwood.

All from A. Thevet, *Les singularitez de la France Antarctique.* Antwerp, 1558. (Catalogue item number 13.)

Thevet describes at some length the people, Tupinambá Indians, their way of life and their foods. Cannibalism is not emphasized and one comes from reading Thevet with a feeling for what would become the "noble savage." He mentions, with some description, many trees which he knows only by their Tupi names. One, which Thevet called "*ahouai,*" is now known to science as *Thevetia ahouai,* to commemorate our author. Thevet gives, too, a good description of the slash-and-burn agriculture used for sweet potatoes and for cassava. He explains how cassava is grated and prepared and remarks that, apart from Peru, Canada, and Florida, all America uses this root for its bread.

Of especial interest is Thevet's discussion of tobacco for which he used the name *petun*. The knowledge of smoking tobacco was not new to Europeans, but Thevet gives for his time the best description of how it was done. Dried leaves were rolled in a palm leaf and the smoke inhaled through the nose and the mouth. The Tupinambá regarded *petun* as a sovereign remedy and knew that it could help one to ignore hunger and fatigue, although Thevet acknowledges that overindulgence could make one nearly intoxicated.

On his return to France, Thevet was appointed to several official positions, and was royal cosmographer to four kings, Henri II, François II, Charles IX, and Henri IV. After Brazil he did not travel again, but in 1575 published *La cosmographie universelle* in which he treats the entire world. Among his comtemporaries Thevet had a reputation for, at the least, exaggeration. As one example out of many that might be noted, in *La cosmographie* Thevet claims that it was he, not Jean Nicot, who first brought tobacco seed to France and refers slightingly to Nicot as one who never travelled. In *Les singularitez* there is no mention of his having brought seed back from Brazil. He might have done so, but his contemporaries regarded his later claim as simply an untruth.

In spite of this reputation, Thevet was for the following ninety years the author most often cited for information on plants of Brazil.

14   CHARLES ESTIENNE (1504?–62)

L'agriculture et maison rustique.

Paris: J. DuPuys, 1567.
The Houghton Library. Bought from the income of the fund bequeathed by Peter Paul Francis Degrand (1787–1855) of Boston.

Estienne, second generation member of a family of scholar-printers, was trained as a physician and came late, through family necessity, to printing. Interested in plants and horticulture, as well as in antiquarian studies, Estienne started in 1535 a series of books for children on topics such as Roman dress, plants, fruit trees, grape-growing and plant nomenclature.

His two great works are a book on dissection and the *Praedium Rusticum* (1554), which tells one how to manage a thriftily-run rural household. Estienne himself translated it into French as *L'agriculture et la maison rustique*, first printed in 1564, with added material provided by Jean Liébault, Estienne's brother-in-law. Among the added items are descriptions of two new world species, the turkey, which had already acquired some popularity in France as a less expensive substitute for peafowl, and tobacco.

The turkey is described with accuracy and with near hostility—it is stupid, noisy ("bruit et fureur") when mature, constantly peeping when young, and the flesh is hard to digest. Turkeys should not be allowed to forage in the garden, but are better turned out along the roadside to fend for themselves.

The chapter on tobacco is another matter. The writer says that "Nicotiane," known in France for only a few years, has taken first place among medicinal plants because of its properties, "singulieres et quasidivines." The writer continues, "i'ay bien sçavoir l'histoire entiere, qu'ay entendue d'un mien seigneur, premier auteur, inventeur, & apporteur de ceste herbe en France."

Jean Nicot, for whom the genus *Nicotiana* is named, was from 1558 to 1560 ambassador from France to Portugal. In Lisbon he was introduced to this foreign plant recently brought from Florida and discovered that it could be used to cure ulcers and wounds. A young page in Nicot's household cured an ulcer on a friend's leg by applying poultices of tobacco. A careless cook nearly severed his thumb with a kitchen knife, but the steward applied nicotiane and it healed well. A woman who had on her face a severe rash was cured in eight to ten days of tobacco application and about this time people in Lisbon started to call this marvelous herb the "ambassador's plant."

A good description of the plant is given with the observant comment that the seeds are the smallest in the world—not strictly true, but they would have seemed so to Nicot. Then instructions follow on sowing of the seeds and on later transplanting of the seedlings—just as tobacco is grown today. In all of this discussion there is no mention made of tobacco being smoked for pleasure—only its medicinal uses. The writer ends by saying that all the above is the true story, told to him by Nicot himself.

Rath, Gernot. "Charles Estienne: Contemporary of Vesalius." *Medical History* 8: 354–359. 1964.

Renouard, Antoine Augustin. *Annales de l'imprimerie des Estienne.* 2d ed. Paris: J. Renouard, 1843.

15   JEAN DE LÉRY (1534–1613)

Histoire d'un voyage faict en la terre du Brésil.

Geneva: A. Chuppin, 1578.
The Houghton Library.

The short-lived (1555–60) Huguenot colony established at Guanabara Bay by Nicolas Durand de Ville-gagnon generated two very popular books which dealt with the native people, the exotically cannibal Tupinambá, and the country in which they lived. André Thevet, a Franciscan who spent a few months in the colony, was in print in 1557 with *Les singularitez de la France Antarctique.* In 1575 Thevet published *La cosmographie universelle* which repeats material drawn from his earlier books but with a number of significant additions. In one of these Thevet accused a group of Calvinist ministers, sent from Geneva to the colony, of treason and laid upon them the blame for the failure of the colony.

This brought a rapid response from Jean de Léry, a Burgundian who had studied in Geneva and who was one of the accused missionaries. One of de Léry's goals was to set straight the record of *Les singularitez* and to rebut statements made by Thevet which he regarded as exaggeration or even as lies. He also presents a great deal of information about the part of Brazil which he saw and the chapter in which de Léry discusses "all those things generally said to have a vegetative soul" is excellent. His descriptions of the plants are clear enough that many of the species can be identified.

The first plant discussed is brazilwood, *Caesalpinia echinata,* the valued red dye-wood which was at that time still the reason for European settlements in Brazil. De Léry tells the fine story of the colonist who wished to bleach everyone's shirts by adding wood ash to the washing lye. Of course it was ash of brazilwood and the colony thereafter wore indelible pink. We find, too, a fruit which the Tupi call *manobi.* De Léry says that these fruit are connected to each other by little filaments and grow under ground like truffles. This is description enough to identify *Arachis hypogaea,* the peanut.

Everyone who went to Brazil was impressed by the use of tobacco which the Tupi called *petun* and used for ceremonial purposes and to ward off hunger. Most Europeans, as did de Léry, tried it and found that it did seem to be an appetite suppressant. Thevet in 1557 had also described tobacco and its uses, but in his *La cosmographie* of 1575 Thevet claimed that it was he, not Jean Nicot, the French ambassador to Portugal in 1558–60, who had brought seeds to France and introduced tobacco. For this de Léry sharply criticizes Thevet and effectively calls him a liar. True or not, the genus, *Nicotiana,* to which tobacco belongs bears Nicot's name as does the alkaloid—nicotine.

When the information on plants in de Léry's book is compared with that given by Thevet, it is clear that de Léry's is at least as good, and for some things, better, but until the middle of the next century it was Thevet's book which generally was cited as the definite word on Brazilian plants.

Hemming, John. *Red Gold: The Conquest of the Brazilian Indians.* Cambridge, Massachusetts: Harvard University Press, 1978.

Léry, Jean de. *History of a Voyage to the Land of Brazil.* Translated by Janet Whatley. Berkeley: University of California Press, 1990.

Reverdin, Olivier. *Quatorze Calvinistes chez les Topinambous: Histoire d'une mission génèvoise au Brésil, 1556–1558.* Geneva: Droz, 1957.

16 WILLEM PIES [PISO] (1611–78)
GEORG MARCGRAVE (1610–44)

Historia naturalis Brasiliae.

Leyden: F. Hack, 1648; Amsterdam: Elzevir, 1648.
The Houghton Library. Gift of Daniel B. Fearing, Class of 1882.

The Dutch West India Company was established in 1621 just as a twelve-year truce between the Dutch Republic and Spain was broken. The Company's charge was twofold—to regulate the trade illicitly carried on with Spanish and Portuguese possessions in America and the west coast of Africa, and to establish settlements in areas where there seemed likely to be products of commercial value.

As Spain and Portugal were then politically united, an attack on a vulnerable Portuguese colony was effectively an attack on Spain. The Dutch first attacked Bahia and briefly held it, but were re-

Boiling down of sugar syrup and molding of sugar loaves. From W. Pies and G. Marcgrave, *Historia naturalis Brasiliae.* Leyden and Amsterdam, 1648. (Catalogue item number 16.)

pulsed, so the Company then turned towards Pernambuco, economically the most important part of Brazil because of its cane growing and sugar manufacture.

The first Dutch settlement was made in 1630 and six years later the Company appointed Count Johan Maurits of Nassau-Siegen as governor-general of the colony. Count Johan Maurits' capital was at Recife, and with the intention of making this part of America known to Europe he brought to his court artists and scholars. Willem Pies was Johan Maurits' personal physician and was responsible for the Count's botanical garden. Georg Marcgrave, a German who knew Jan de Laet, a director of the West India Company, was appointed astronomer and geographer.

Nassau-Seigen's remarkable court lasted for only eight years; in 1644 the Company recalled the Count to the Netherlands and he took with him the great mass of natural history specimens collected by Pies and Marcgrave. The Count and Pies returned to successful careers at home, but Marcgrave, by far the better natural historian of the two, was sent to Angola where he died soon after his arrival.

Pies, the physician, had concentrated on plants of medicinal value, but Marcgrave traveled widely in northeastern Brazil during his six years there (1638–44), and had amassed notes, illustrations and actual specimens of dried plants. Most of the natural history materials were turned over for editing to Jan de Laet who had a strong interest in the subject.

De Laet worked rapidly and surprisingly well from Marcgrave's unorganized notes which he found, to his surprise, were written in code, apparently to prevent anyone else—Pies, most likely—from using them if Marcgrave could not. Presumably Johan Maurits held the key and the *Historia naturalis Brasiliae* was printed early in 1648.

Pies' sections on diseases and medicinal plants come first (132 pages) followed by Marcgrave's natural history (303 pages) in eight books, the first three dealing with plants. It is a magnificent work which points the way for all later books on natural history which deal exhaustively with a particular region, and it is the first of its kind for the new world. There would be no equivalent collection of

Grinding of cassava, *Manihot esculenta*. From Pies and Marcgrave, *Historia naturalis Brasiliae*. Leyden and Amsterdam, 1648. (Catalogue item number 16.)

natural history information for the new world until the great Spanish expeditions of the late eighteenth century.

Ten years later Pies published a "second edition," *De Indiae utriusque re naturali et medica libri quatuordecim* in which he appropriated much of Marcgrave's material and used it with an inexcusable minimum of acknowledgement. In the world of systematics such deeds need not go unpunished. In 1737 Linnaeus, who was always one to speak frankly, commemorated Piso by naming for him the genus *Pisonia*. Linnaeus remarked that *Pisonia* is a horrid tree with spines, as horrid as the man himself, if what Marcgrave's brother said of him were true.

Whitehead, P.J.P. "Georg Markgraf and Brazilian Zoology." In *Johan Maurits van Nassau-Siegen, 1604–1679: A Humanist Prince in Europe and Brazil,* edited by E. van den Boogaart. The Hague: The Johan Maurits van Nassau Stichting, 1979.

17   Francisco Hernández (1517–87)

Rerum medicarum novae Hispaniae…historia.

Rome: V. Mascardi, 1651.
Library of the Gray Herbarium.

Hernández, born near Tolédo, was the first of the great naturalists of Spain. Educated at Salamanca where he took a degree in medicine, Hernández became physician to Philip II. Philip was interested in, among other things, animals and plants and there were royal gardens at El Escorial, at Segovia, and at Aranjuez. Perhaps he realized that the silver and gold from the new world might eventually be outweighed by the value of its biological resources.

Guaiacum from the West Indies had for many years been imported to Spain and then exported throughout Europe as the leading cure for syphilis. However this valuable commodity was slowly being replaced by other drugs and, perhaps, there could be found in the new world other products of similar value. In 1570 Philip sent Hernández and his son, funded with 60,000 ducats, to Mexico with the charge to produce an inventory of the antiquities and of the natural products of Mexico.

They went expecting to spend five years; in fact, they stayed for seven, travelling through the country with native artists, interviewing native physicians and herbalists, examining manuscripts, and col-

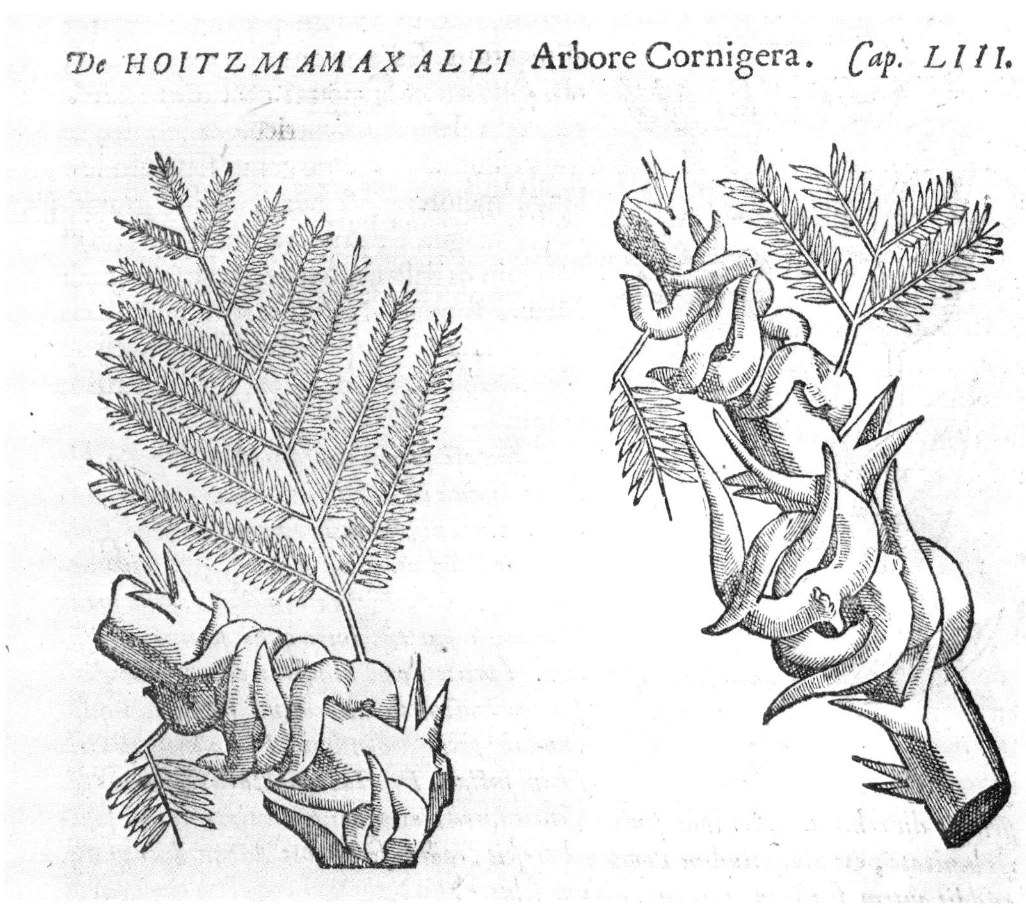

Thorns of the bull's-horn acacia, *Acacia cornigera*. From F. Hernández, *Rerum medicarum Novae Hispaniae…historia*. Rome, 1651. (Catalogue item number 17.)

lecting plants for trial use in local hospitals. In 1577 the Hernández' returned, with a mass of information, to Spain. It was, however, not published, most likely because of its size, and was bound up and placed in the library of El Escorial.

After the death of Hernández in 1587, Leonardo Recchi, who was also one of Philip's physicians, made from the whole work an extract which also was not published. The manuscript was again extracted by Juan Nieremberg and this extract was printed in 1635 as *Historiae naturae maxime peregrinae*. The original manuscript was destroyed in the Escorial fire of 1671, and only the extracts remain.

That made by Recchi was sold early in the seventeenth century to the Accademei dei Lincei and a few of the Lincei prepared it for publication, often adding extensive notes. A few copies were printed in 1628, but the main printing was completed only in 1651 thanks to a subvention from the Spanish government. Even for 1651 *Rerum medicarum novae Hispaniae historia* is a remarkably wide-ranging work. In 1648 had been printed the *Historia naturalis Brasiliae* of Pies and Marcgrave so that within these few years the doors to an understanding of the natural history of tropical America had been opened wide. Had Hernández' book been printed in the 1580's it would have been enormously influential.

The pictures are redrawn from the originals of Hernández' Aztec artists and they vary greatly in quality. Some at once confirm the identity of the plant depicted; others are frustratingly unidentifiable. On page 86, under the heading "De Hoitzmamaxalli Arbore Cornigera," we find a biological first. The image shows a plant with greatly enlarged thorns, in shape much like the horns of a bull. Hernández says that within these thorns are "generated" ants which give painful bites. We find here the first mention in print of an ant-plant relationship. The plant is *Acacia cornigera*, native to Mexico, and the thorns house nests of the ant *Pseudomyrmex ferruginea* F. Smith. The ants are sheltered in the thorns and, in return, serve as aggressive deterrents to herbivores.

Janzen, D.H. "Interaction of the Bull's-horn Acacia *(Acacia cornigera* L.*)* with an ant inhabitant *(Pseudomyrmex ferruginea* F. Smith*)* in Eastern Mexico." *University of Kansas Science Bulletin* 67: 315–558. 1967.

18 OTTO BRUNFELS (1489–1534)

Herbarum vivae eicones.

Strasbourg: J. Schott, 1530.
The Houghton Library. From the Collection of Philip Hofer, Class of 1921, for the Department of Printing and Graphic Arts.

Early in the sixteenth century European botany was chiefly concerned with the identification and correction of errors which had crept into the work of Dioscorides. About A.D. 60, Dioscorides had written in Greek a pharmacopoea, best known by the Latin title *De materia medica,* which was highly regarded by Galen and which is the ancestor, by various routes, of late-classical and medieval herbals.

After the collapse of the Roman empire, the Greek text of Dioscorides remained known in the Byzantine world, but in the west Dioscorides was known only in Latin translation—sometimes much modified or with material added, with many of the additions coming from Arab authors. However, in the middle of the fifteenth century Greek manuscripts of Dioscorides became available in Italy and there were soon printed editions in Latin in 1478 and in Greek in 1499.

It was then possible to return to the original Greek, to make corrections, and to strip away the overlay added by later authors. The goal was a purified Dioscorides, which would provide a baseline against which the plants used as medicinal simples could be compared. In the sixteenth century came

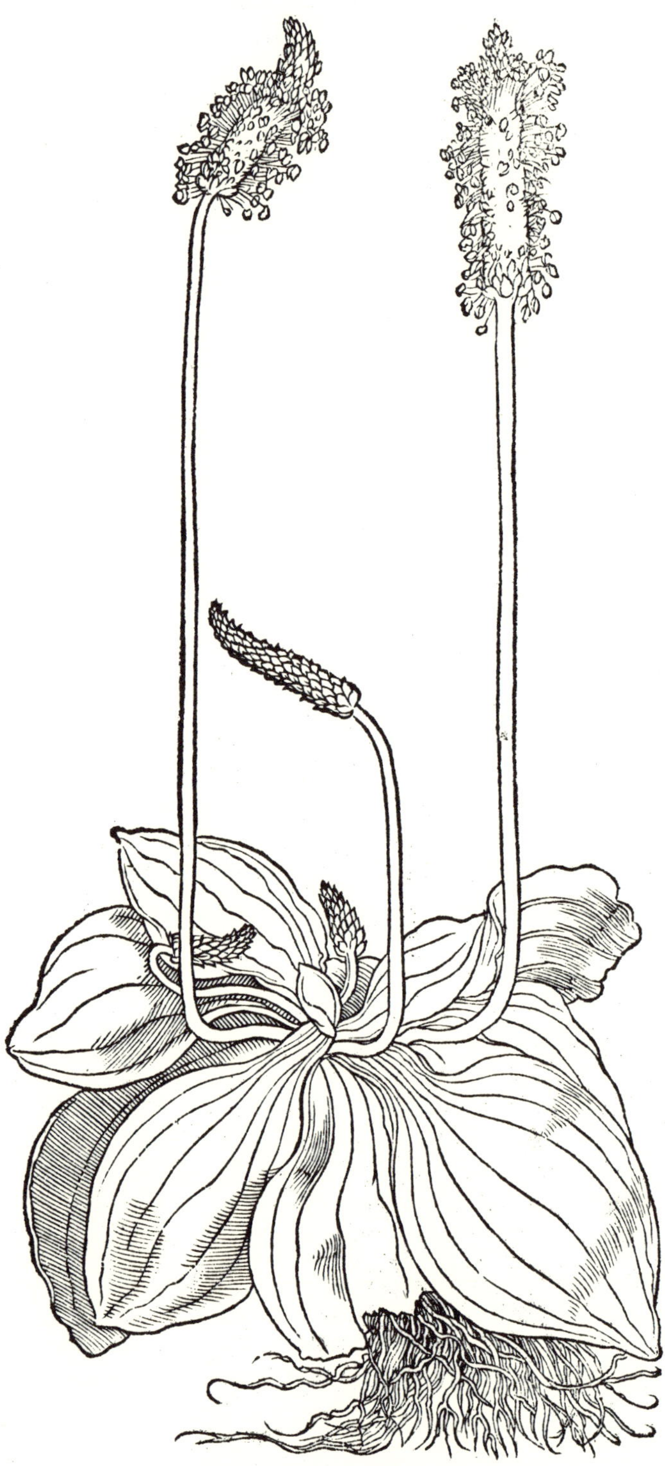

Plantago major, one of the commonest of introduced weeds; called "white man's foot" by New England Indians. From O. Brunfels, *Herbarum vivae eicones*. Strasbourg, 1530. (Catalogue item number 18.)

a parade of works which, in English, are generally referred to as "herbals." Most of them are commentaries on Dioscorides, written in the hope that unambiguous links could be made between the medicinal plants available to European physicians and those discussed by Dioscorides. The attempt was doomed to only a moderate degree of success, for many of the Mediterranean plants mentioned by Dioscorides simply do not occur in trans-alpine Europe.

The first of these was the *Herbarum vivae eicones* of Otto Brunfels. Brunfels, born in Mainz, was variously a Carthusian monk, a schoolmaster, and finally town physician in Berne where he died. The lifelike illustrations, the first of their kind, stand in sharp contrast to the crudely stylized images found in most—but not all—contemporary printed books about plants. However, the text looks to the past and consists of a concatenation of quotations from earlier authors, put together in an attempt to equate German plant names used by druggists with those in Dioscorides.

Brunfels' book is concerned with plants growing around Strasbourg, and we should not expect to find any mention of new world plants, nor do we. However, in the introductory "Encomium medicinae," a history of medicine, we find mentioned on page fifteen Ulrich von Hutten, a German humanist who had struggled for years with the physically devastating mercury treatments used to treat syphilis. In desperation Hutten tried guaiacum which had recently been introduced to Europe and believed himself cured. To let the world know of the new drug Hutten wrote a little book, *De guaici medicine et morbo gallico,* printed in 1519. It was immediately translated into German (1519) and into French (1520), and Brunfels thought it important enough to be included in his history of medicine. Such references to cures for syphilis emphasize the impact that the disease had on Europe. At the time Brunfels' book was printed, syphilis had been in Europe for thirty-five years, and was still new and devastating. The most used cure was guaiacum, the first plant product of the new world to become important in European medicine.

Arber, Agnes. "The Draughtsman of the *Herbarum vivae eicones.*" *The Journal of Botany, British and Foreign* 59: 131–132. 1921.

Dilg, P. "Die botanische Kommentarliteratur in Italien um 1500 und ihr Einfluß auf Deutschland." In A. Back & O. Herding (eds.), *Der Kommentar in der Renaissance.* Boppard: Boldt, 1975.

Jacobs, M. "Revolutions in Plant Description." *Landbouwhogeschool Wageningen Miscellaneous Papers* 19:155–181. 1980.

Schipperges, H. "Ideologie und Historiographie des Arabismus." *Sudhoffs Archiv Beiheft* 14–26. 1961.

19 ANTONIUS MUSA BRASAVOLA (1500–55)

Examen omnium simplicium medicamentorum.

Lyon: J. Barbou, 1537.
Library of the Arnold Arboretum. The gift of Sarah C. Sears.

Antonius Brasavola of Ferrara was a physician, said to have treated more than 1500 patients in a single year, and a prolific writer. He wrote on medical and theological subjects, compiled an index to the works of Galen, and published a Greek grammar. *Examen omnium simplicium* was written to address a serious problem—how to equate the plants sold in the apothecary's shop with those described by Dioscorides.

It is presented as a dialogue between a bright young physician, named Brasavola, and an ancient and crotchety druggist, Senex. Senex and his assistant, Herbarius, one day meet Brasavola while all three are out botanizing in the hills around Ferrara. Senex is an obstinate old fellow who "knows" medicinal plants, but by the end of the day's outing, Brasavola has convinced him that bright young men with a firm base in Dioscorides and Pliny can know things too, and can know them better than can Senex.

The *Examen* is full of comments about ignorant and careless apothecaries, and even about ignorant physicians. Those who are secure in their ignorance and arrogantly refuse to learn, are, Brasavola says, a miserable lot. He does admit that learning medicinal plants, simples, is not easy, for there are drugs used today (1536) which were not known to Dioscorides. This is, in part, a consequence of voyages of exploration. In fact, Brasavola says, and this is indeed remarkable for a physician of his time, that Dioscorides, Pliny, and Theophrastus did not know a hundredth part of the herbs in the world: we learn every day of new ones and so medicine advances.

One of these new drugs, an important one, is guaiacum, *Guaiacum officinale,* not known to Dioscorides, but as Brasavola remarks, this is not surprising, for the very lands in which it grows, were not known to Ptolemy and other early geographers.

Guaiacum, in Brasavola's time often called "lignum sanctum," and now known as "lignum vitae," had been known in Europe for about thirty-five years, and was highly regarded (quite wrongly so) as a cure for syphilis. The medicine was prepared by steeping chips of the wood in water which was drunk by the patient. Brasavola points out that even this new drug was already a problem for there were three things sold in druggists' shops as lignum sanctum. One was the true guaiacum, but people often were deceived and sold the wrong wood and Brasavola explains how the color of the heartwood could be used as a guide to the genuine lignum sanctum.

Guaiacum is the only medicinal plant of the new world mentioned by Brasavola. He does note that growing in clay pots in the windows of Ferrara was a plant called commonly "Spanish" or "Calicut pepper," and so we know that American chili peppers were already being grown as ornamentals.

Durling, Richard J. *A Catalogue of Sixteenth-Century Printed Books in the National Library of Medicine.* Bethesda, Maryland: U.S. Department of Health, Education, and Welfare. Public Health Service, 1967. Durling lists twenty works by Brasavola.

Thorndike, Lynn. *A History of Magic and Experimental Science.* Volume V: "The Sixteenth Century." New York: Columbia University Press, 1941. See Chapter XX: "Brasavola and Pharmacy," pp. 445–471.

20   JEAN RUEL (1474–1537)

De natura stirpium libri tres.

Basel: Froben Office, 1537.
Library of the Arnold Arboretum. The gift of Francis Skinner of Dedham.

Ruel, who was for several years physician to Francis I of France, is best known for his translation into Latin of Dioscorides' *De materia medica* (1516), recognized for years as the best of several done at much the same time, and for *De natura stirpium,* first printed at Paris in 1536. It is not a herbal, but is a solid treatment of botany in which Ruel describes plants and their parts. Medicinal properties are mentioned, but they are not the most important part of Ruel's treatment.

Although 1536 is early indeed to find new world plants mentioned in a book of this kind, Ruel has a section on guaiacum. Syphilis had spread so rapidly and with such virulence through Europe that even if no other American plant had been noticed in a book on plants or on medicinal simples, guaiacum would have been. As a cure for syphilis, guaiacum is useless, a fact that would eventually be realized, but at this time a new world plant was still thought to be the cure for a new world disease. Forty years after syphilis had appeared in Europe, Ruel emphasizes the newness of the disease: "novum morbum, omnique aevo priore incognitum; qui per Hispanias, Gallias, Italiam, totamque Europam vagetur."

In this same section Ruel describes brazilwood. "Bersilicum vocant, vel potius bresilum," and explains its use as a dye-wood. He mentions, too, a problem already apparent; the stuff wasn't colorfast and cloth dyed with brazil tended to fade from its initial highly-prized red to a less desirable pink.

In Book II, Ruel has a series of short chapters on grasses. In Chapter 27, headed "milium," that is, millet, is an intriguing note about "milium saracenium" which, Ruel says, has been known in France for only about fifteen years, but is now widely grown in gardens: "Hodie Galli in hortis ostentationis gratia fereunt." This Saracen millet, which is five to six feet high and has grains the size of peas, is, most likely, maize, although one cannot be sure. There are near-eastern species of the genus *Sorghum* which were sometimes confused with the new world plant. In habit, they can look remarkably like maize although they do not, of course, have the unique husk-wrapped ears of the new world plant. Still, it is a pleasant thought that some people in the France of Francis I might have been growing ornamental corn.

The scientific name for maize, new world corn, is *Zea mays,* so named by Linnaeus. At Ruel's time, and until the publication in 1737 of Linnaeus' *Genera Plantarum,* the word "zea" did not mean maize. "Zea" or "zeia" is used by Homer, by Hesiod, and by later Greek writers, to mean *Triticum monococcum,* a close relative of wheat (in English called "spelt") and it meant spelt for nearly 2500 years until Linnaeus' wholesale disruption of plant names.

21   LEONHARD FUCHS (1501–66)

De historia stirpium commentarii insignes.

Basel: M. Isingrin, 1542.
The Houghton Library. Bequest of William King Richardson, Class of 1880.

De historia stirpium is the second, after Brunfels' *Herbarum vivae eicones* (1530–32), of the beautifully illustrated herbals of the sixteenth century. The quality of the images is such that, in many cases, the plant shown can be identified to species. Between Brunfels and Fuchs, Jerome Bock (Hieronymus Tragus) published in 1539 *New Kreuetter Buoch,* with good descriptions of the plants, better, perhaps, than in either Brunfels or Fuchs. However, there are no illustrations and the first edition of Bock's book had little impact. The images in Fuchs' *De historia stirpium* are good; so good, indeed, that they tempt one to add color and few copies are without at least a bit of added color.

Fuchs, born in Bavaria, received a degree in medicine in 1524 from the University of Ingolstadt. After several years of teaching there, and a few years as physician to the Markgraf of Brandenburg, Fuchs went in 1535 as Professor of Medicine to Tübingen, and there he spent the rest of his life. He was a prolific author, producing in addition to the herbal and its later editions, commentaries, translations, medical works and pamphlets on religious matters, for Fuchs had become an ardent Protestant.

In *De historia stirpium* Fuchs' concern was to clean up pharmacy in Germany—to identify the plants described by Dioscorides, Pliny, and Galen—and to provide an accurate image of each. Pharmacy was in terrible condition. In his "Dedicatory Epistle" Fuchs complains that physicians feel it beneath themselves to know anything about plants and depend upon druggists for accurate identifications. "And so it comes to pass that the druggists—God knows that they themselves are for the most part an illiterate bunch—leave this all to foolish and superstitious old women who gather herbs and roots." It all reminds one of the earlier jibe of Erasmus about the doctors at Paris who could not identify the parsley in their salads.

Fuchs provided to medical students beautifully done pictures of plants, made to his specifications, and he kept control over the artists. In the preface he states that he had not let shadowing be used, and that he had not allowed the craftsmen *("artifices")* to indulge their artistic fancies, lest these fancies cause the drawings to be inaccurate.

Among the plants of trans-alpine Europe there are four from the new world: the first images of chilies, and of maize, new world beans, and squash. Fuchs remarks that the chilies and squashes had been recently introduced to Germany. He says nothing about the origin of the beans, but the plant shown is clearly *Phaseolus vulgaris* from America. Maize, Fuchs calls "*Turcicum-frumentum,* Türkisch Korn" and notes that it was recently introduced, "it is said," from Asia and Greece, but is now fairly common and is grown in many gardens.

The same error about the homeland of maize appears in herbals until 1570 when Pierandrea Mattioli suggested that it is, in fact, an American. Mattioli's text strongly suggests that he had read Oviedo's *Historia general y natural de las Indias* where he explains maize and its use in detail. The mistake about origin is understandable. In Spain maize had been grown for many years and Oviedo reported that in 1530 he had seen growing at Avila a plot of maize about ten hands [80 inches] high, as good as any he had seen in America. But plants can be so easily carried from place to place by a few seeds sent to a correspondent who the next year sends his seeds along to another, and knowledge of the true home of the species is soon lost.

Finan, J.J. "Maize in the Great Herbals." *Ann. Missouri Bot. Gard.* 35:149–191. 1948. Useful, but to be used with some caution.

Sprague, T.A. & E. Nelmes. "The Herbal of Leonhart Fuchs." *Journ. Linn. Soc. London* 48: 545–642. 1931.

22   Conrad Gesner (1516–65)

Horti Germaniae.

Strasbourg: J. Rihel, 1560.
Library of the Arnold Arboretum. Gift of Sarah C. Sears.

Conrad Gesner (or Gessner) had a breadth and diversity of interests and knowledge which fit with difficulty into our modern pigeonholes of classification. Born in Zurich, and educated at Bourges, Strasbourg, and Paris, Gesner received his degree in medicine at Basel and returned to Zurich. Here he spent most of the rest of his life teaching at a theological seminary. Apart from this work Gesner wrote—and wrote—on subjects from theology to zoology. By those who would assign paternities, Gesner has been called the father of both bibliography and of veterinary medicine.

He is best known for his great *Bibliotheca universalis* which contains annotated entries for nearly 12,000 works on all subjects except medicine for which, Gesner apologizes, his material was incomplete. Gesner's other major work is *Historia animalium* which amounts to more than 4000 pages. Plants, too, interested him; he planned a great botanical encyclopaedia which remained unpublished until the middle of the eighteenth century. His interest in plants was practical as well as theoretical, for

Earliest printed image of *Zea mays*. From L. Fuchs, *De historia stirpium*. Basel, 1542. (Catalogue item number 21.)

he refers to himself as "Con. Gesnerus cuius hortulus quidem perangustus est," or "Conrad Gesner, whose small garden is narrow indeed."

About 1560, Gesner was asked to edit and see through the press some of the works of Valerius Cordus, a brilliant young botanist who had died in 1544. This he did and printed with Cordus' *Annotationes* his own *Horti Germaniae*. Here Gesner lists in alphabetical sequence the plants grown in his own garden and in those of several of his many correspondents.

A surprising number of the plants are Americans. There is maize, of course, which Gesner describes well, though he does refer to the "useless" flowers at the top of the stem. In fact, the flowers of the tassel are staminate and produce pollen. Gesner himself did not have maize in the "hortulus perangustus"—too large a plant, one wonders?—but he did have "flos indicus" (marigolds, native to Mexico), "pomum aureum vel amoris" (tomatoes, which he had tried: "nec insuavis, nec noxius in cibo") and "piper indicum," chili peppers of several kinds.

Gesner had in his garden another, and unexpected, new world plant, "ficus indica, quam aliqui ridicule Opuntiam nominant." This is a cactus, the prickly pear, some species of the genus *Opuntia*. Gesner's "ridicule" refers to Pliny's use of the name "opuntia" for some spiny plant of the old world; to apply it to a plant of the new world was clearly absurd. He remarks that north of the Alps and in northern Italy *ficus indica* grows well, but does not produce seed. South of the latitude of Rome and in Greece *Opuntia* did set seed and we see here the first hint of a new world plant which soon became naturalized in Europe and in some places became a troublesome weed.

One hundred thirty years later, Dr. Tancred Robinson wrote to John Ray in Cambridge, England, that he had seen *ficus indica* growing commonly near Rome and wondered if it were native there. Ray replied, "though the *Ficus indica* be so frequently found growing plentifully in Italy, that one would be apt to think it a native of that country, yet doubtless it is originally a stranger and an American."

Lankester, Edwin, ed. *Correspondence of John Ray.* London: The Ray Society, 1848.

Wellisch, Hans H. *Conrad Gessner: A Bio-Bibliography.* Zug, Switzerland: IDC AG, 1984.

23   MATTHIAS DE L'OBEL (1538–1616)

Plantarum seu stirpium historia.

Antwerp: C. Plantin, 1576.
Library of the Massachusetts Professorship of Natural History. Deposited temporarily in the Library of the Arnold Arboretum, March 9, 1894.

During the last thirty years of the sixteenth century, European botany lay in the hands of three botanists of the Low Countries, Rembert Dodoens of Mechelen, Charles de l'Escluse (Clusius) of Arras, and Matthias de l'Obel, born in Lille. They shared ideas, they shared illustrations of plants, and they had in common a remarkable printer, Christopher Plantin of Antwerp, who spread their books and ideas across Europe.

L'Obel was a physician who had a degree in medicine from Montpellier. In 1569 he and Pierre Pena, a physician friend from Montpellier, went to England—in large part for reasons of religion, for both were Protestants. They brought along a manuscript which was printed in 1571 by Thomas Purfoot of London as *Stirpium adversaria nova, (New Notes on Plants)*. Purfoot seems to have overestimated the demand for the book, for Plantin bought from him 800 unbound copies which he used in 1576 as part of a larger work, *Plantarum seu stirpium historia,* for which l'Obel had provided much new material.

By 1570 there were perhaps ten new world species which were becoming familiar in Europe to educated people who had some knowledge of plants as medicinal simples and to people who were able to have in their gardens rare and unusual plants. Perhaps a dozen more were known by name and by reputation, all of them from Spanish America.

L'Obel describes at least fifteen American plants, maize, tomatoes, chilies, sunflowers, nasturtiums, sweet potatoes, marigolds, tobacco, kidney beans, and several others used in medicine. For l'Obel, tobacco is still a plant with medicinal virtues, although he provides the first image in a herbal of a person smoking.

Next to l'Obel's picture of the tobacco plant is a drawing of a man's head with a smoking "cigar" in the lips. L'Obel describes the cigar as dried tobacco wrapped in a palm leaf, and remarks, "nicotiana inserta infundibulo ex quo hauriunt fumu Indi et naucleri" ("nicotiana inserted into a tube from which Indians and sailors suck in smoke").

For new world beans l'Obel can cite common names, "kidney beans" in English and in German, "Welsch [Welsh, i.e., exotically foreign] bonen;" kidney beans or string beans, etc., all *Phaseolus vulgaris,* had rapidly become popular in Europe for they were thought to have a better flavor than did *Vicia faba,* the familiar broad or fava beans of the old world. Among other advantages they had fewer of the complex carbohydrates which make fava beans difficult to digest. And in Mediterranean populations there is a comparatively high frequency of a genetic disorder, favism, which often is manifested as hemolytic anemia. When sufferers from favism ingest broad beans, the anemia is aggravated, but new world beans of the genus *Phaseolus* lack the compounds responsible for this, and they rapidly became popular and commonly grown.

Louis, A. *Mathieu de l'Obel 1538–1616: Épisode de l'histoire de la botanique.* Ghent-Louvain: Story-Scientia, 1980.

24   NICOLÁS MONARDES (ca. 1493–1578)

Primera y segunda & tercera partes de la historia medicinal…

Seville: A. Escrivano, 1574.
The Houghton Library. Gift of Samuel A. Eliot.

We know little of Nicolás Monardes other than that he received a degree in medicine from Alcalá de Henares and became a successful and respected physician in Seville. Although Monardes never visited the new world he was well-positioned in Seville to meet those returning and to become familiar with medicinal plants they brought.

In 1569 Monardes published *Dos libros, el uno que trata de todas las cosas que traende de nuestras Indias Occidentales, que sirven al uso de la Medicina;* the second of *Dos libros* treats the bezoar stone (a popular nostrum) and a plant called scorzonera. Two years later Monardes published a "Second Part" in which he discusses tobacco and sassafras, both of which had recently come to be used in medicine.

These publications of 1569 and 1571 were so popular that in 1573 Monardes combined them, and with some added material they make up this present book. Had Monardes' book remained only in Spanish, it might have been less well known outside the Iberian peninsula, but Charles de l'Escluse (Clusius) translated it in an epitomized form into Latin and this was several times reprinted. John

First image of *Sassafras albidum*. From N. Monardes, *Primera y segunda...partes de la historia medicinal.* Seville, 1574. (Catalogue item number 24.)

Frampton, an English merchant who had lived in Spain, translated Monardes' book of 1574 into English under the attractive title *Joyfull Newes out of the Newe Founde Worlde.*

Some have found it easy to scoff at Monardes because of his faith in such things as the bezoar stone, but he should rather be admired for his willingness to listen to others and to try new things: he tells us that in his forty years of practice he has made a point of trying new drugs from the West Indies. Monardes remarks that the three chief remedies brought from America are guaiacum, china, and sassafras. Guaiacum was well-known, for a decoction of the wood of *Guaiacum officinale* had been used as a remedy for syphilis since the first years of the century.

China refers to the tuberous rootstocks of species of *Smilax,* a genus of climbing vines in the lily family, which is common enough in eastern North America as the troublesome catbriar or greenbriar. It had first been brought into Europe by the Portuguese, supposedly from China, and when the Spanish found similar species in Mexico, the name was adopted. Sarsaparilla also refers to roots of species of *Smilax.* All were used as sudorifics with the idea that illness could be sweated out of the patient.

In the Second Part, printed in 1571, Monardes introduces sassafras *(Sassafras albidum),* and devotes twenty pages to it. Sassafras came into Spain during the 1560's from Florida and was enthusiastically recommended by Monardes as a near-panacea. He had first heard of it only three years before when a Frenchman told him of its use in Florida. At first, Monardes ignored this recommendation from an amateur, but did at last try it and found that sassafras cured a great many diseases and ailments. The translations by Clusius and by Frampton made Monardes' work known throughout Europe and sassafras was, until the middle of the next century, a desirable commodity which could be brought home from America and readily sold.

25   JOHN FRAMPTON (fl. 1577–96)

Joyfull Newes out of the Newe Founde Worlde.

London: E. Allde, 1596.
The Houghton Library.

John Frampton had lived in Spain as a merchant and found on his return to England that he had time on his hands. Being not pressed with "the former toiles of my old trade, I to passe the tyme to some benefite of my countrie, and to avoyde idlenesse: tooke in hand to translate out of Spanishe into Englishe, the three bookes of Doctour Monardes of Sevill." Frampton indeed made good use of his time, for he also translated into English the travels of Marco Polo; part of Enciso's *Suma de geografía;* and Escalante's work on China.

Frampton continues that Monardes' book on medicinal products from America might be of value to his countrymen for it told of "wonderfull cures of sundrie great diseases, that otherwise…were incurable." That may or may not have been the case, but Frampton's *Joyfull Newes* was the book consulted or carried by English explorers until after the colonization of Virginia, for it was the only accessible source of information on plants of the new world. Frampton translated Monardes' book of 1574 which combined *Dos libros,* printed at Seville in 1569, and *Segunda parte del libro…* of 1571. The translation was first printed in 1577, with a second edition in 1580, and this third in 1596.

Monardes has a long treatment of tobacco at the beginning of his second book; i.e., that part first printed in 1571. These remarks are especially interesting for he (Monardes) says that tobacco had been brought to Spain during the past few years, but as an ornamental plant rather than for its therapeutic value. However it is now (1571) used more for its virtues than for its beauty. Tobacco is described as being good for headaches, ulcers, wounds, and many other problems which Monardes discusses at some length. It seems almost a panacea, for on a single page it is recommended for worms, chilblains, and toothache, and is especially favored for use on wounds. But there is no mention of how an ornamental plant was found to have such healing properties.

However, in Frampton's *Joyfull Newes* the discussion of tobacco is followed by Frampton's only addition to Monardes' text. This is a translation, essentially word for word, of the chapter on nicotiane, tobacco, in Charles Estienne's *L'agriculture et maison rustique* of 1564, where the story of the discovery of its medicinal uses is told. Tobacco's supposed curative properties were discovered in Lisbon by Jean Nicot who was ambassador from France to Portugal in 1558–60. Monardes surely knew all this but his reticence may have been his Spanish patriotism, reluctant to give credit to anything which came out of Portugal.

Thomas Hariot who describes "Virginia" (in fact, North Carolina) in his *Briefe and True Relation* (1588) refers to Frampton. The plant of interest is sassafras, introduced to European medicine by Monardes, which was for many years a valued article of commerce. Indeed Frampton points out in the introductory "epistle dedicatorie" that valuable medicines are brought from the West Indies to Spain, and then to England, a hint that England might find it both profitable and politically expedient to find its own supplies.

26   Jacques Dalechamps (1513–88)

Historia generalis plantarum.

Lyon: G. Rouillé, 1586–87.
Library of the Arnold Arboretum.

Born near Caen in Normandy, Dalechamps took his degree in medicine at Montpellier as a student of Guillaume Rondelet who taught many of the great botanists of the sixteenth century. Dalechamps spent his life at Lyon where he taught, practiced medicine, and did much translating and editing of earlier works—some of these huge ventures, for he translated Pliny, Galen, and Athenaeus.

His own great work is the *Historia,* printed by Rouillé after Dalechamps had become too infirm to see it through the press. Most of the book is Dalechamps,' but over several years he had help from two other of Rondelet's students, Jean Bauhin and Jean Desmoulins.

None of the three is mentioned on the title page, which is a plea from the printer to buy the book, a gigantic thing of more than 1900 pages. Contemporaries criticized the *Historia* for being carelessly put together, and it does show both the absence of a firm hand to direct editing and proofreading, and the presence of too many hands in its compilation. Caspar Bauhin, professor of medicine at Basel, commented that although much of the book was necessary for students of botany and medicine, parts of it would frighten them away from these subjects and other parts would mislead them. In this criticism there is no mention that Bauhin's older brother, Jean, had worked on the book.

But to consider the good—the *Historia* was in 1586 the one complete botanical work available, including more than 2700 species. Then, too, we find here the first systematic gathering in a herbal of what was known about non-European plants. In Book 18, "In quo describuntur et ad vivum depinguntur Plantae Peregrinae," are described both American and east Asian species. Dalechamps certainly did not include all of the plants that he might have, but Book 18 gives to the reader a 175-page-long summary of foreign plants.

The first chapter is on guaiacum, the great "cure" for syphilis, and still one of the most important items of foreign *materia medica,* although it had for some years been gradually replaced by sassafras which was introduced to European medicine by Nicolás Monardes of Seville. At the end of Book 18 come two surprises. Neither is absolutely new with Dalechamps, for he borrowed both from Clusius, but there is the first mention in a herbal of coca, with a careful description of the plant and an explanation of its use in Peru. Then there is a very short chapter, "Tubuli ad Asthma utiles," the first notice in a herbal of smoking as a health aid. Tobacco had for twenty-five years been used as a vulnerary in the form of poultices applied to ulcers and wounds, but the smoking of tobacco still was something done by new world Indians and by adventuresome explorers. To see its use recommended by a conservative physician is surprising.

27   Charles de l'Escluse [Clusius] (1526–1609)

Rariorum aliquot stirpium, per Pannoniam, Austriam, et vicinas quasdam provincias observatarum historiae.

Antwerp: C. Plantin, 1583.
Library of the Gray Herbarium. The gift of Asa Gray.

De l'Escluse, better known by his latinized surname of Clusius, is the most engaging figure in sixteenth-century botany. Born to a Protestant family in Arras, Clusius was educated at Louvain as a lawyer, and later studied at Montpellier under Guillaume Rondelet. Possessed of a lively and enquiring mind, Clusius traveled widely in Europe and was a prolific writer. Admired by his contemporaries, Clusius was a central figure in the network of scholars and naturalists, and thus learned of new plants from new worlds almost as they reached Europe. He had, for example, close contacts in London with the men around Francis Drake and Walter Raleigh and through them regularly received plant material from America.

Some of Clusius' most useful work lies in his many translations from the vernacular into Latin, such as Garcia da Orta's work on medicinal plants printed in Portuguese at Goa in 1563. Clusius made a translation, actually an epitomized translation, and had it printed by Plantin in 1567, so making the work widely available, and he did the same for Nicolás Monardes whose book on medicinal plants of Spanish America was first printed in Spanish at Seville.

For this exhibition I have selected not one of the great folios, but a tidy little octavo printed by Plantin in 1583. In 1573 Clusius was invited by the Emperor Maximilian II to Vienna where he was for several years associated with the imperial gardens. Clusius explored the mountains of Austria, Styria, and western Hungary (Pannonia) and this little book describes alpine plants. It is the first work on the natural history of the region and comes close to being, in our modern use of the word, a flora.

Ornamental plants always were of interest to Clusius—he popularized tulips in Europe—and one American species appears. Clusius tells us that "gescheket indianishe blumen" are now popular in Vienna ("in deliciis est"). He remarks that this plant had different names in different places; in Italy it is "Mexican jasmine;" in Spain, "marabillas del Peru." The name and the figure show that it is most likely *Mirabilis jalapa,* still known in English as "marvel of Peru" or "four-o'clock," for the flowers open late in the afternoon.

At the end of the book is "Stirpium nomenclator pannicum," which gives Hungarian names for some common plants. Four-o'clocks have no name in Hungarian, but there is *török buza,* Turkish wheat, which may well have been maize, and *török borsó,* Turkish peas, which is *Phaseolus vulgaris,* the new world common or string or runner bean. Ninety years after the discovery a few plants from the new world were common enough in eastern Europe to have been given vernacular names.

Hopper, Florence. "Clusius' World: The Meeting of Science and Art." In L. Tjon Sie Fat and E. de Jong (eds.), *The Authentic Garden.* Leiden: Clusius Foundation, 1991.

Hunger, F.W.T. *Charles de l'Ecluse (Carolus Clusius), Nederlandsch Kruidkundige 1526–1609.* The Hague: Martinus Nijhoff, 1927.

Lenger, Marie-Thérèse (ed.). *Bibliotheca Belgica.* Brussels: Culture et Civilisation, 1964–1975.

28   JACOB DIETRICH [TABERNAEMONTANUS] (1520?–90)

Eicones plantarum.

Frankfurt am Main: N. Bassé, 1590.
Library of the Arnold Arboretum. Gift of Sarah C. Sears.

Jacob Dietrich or Theodore was born and spent most of his life in Bergzabern, and from that town he took his latinized surname, Tabernaemontanus, by which he is generally known. He was a practicing physician who spent his leisure time in preparing a herbal, the *Neuw Kreuterbuch,* which was printed after 30 years' work by his friend, Nicolas Bassé.

Tabernaemontanus is usually presented as a minor figure among sixteenth-century botanists, but he is one of the authors mentioned by Caspar Bauhin, professor of medicine at Basel, as those used by his students. The *Neuw Kreuterbuch* is a huge book which includes, according to the title page, descriptions of more than 3000 plants. It was known and respected for its excellent illustrations which made it useful to medical students.

There are several pages devoted to maize, showing whole plants and individual ears, and there are included at least a dozen other American plants, some of which had by then become quite familiar in Europe. Tabernaemontanus even has the tomato, "Lieboepffle," showing a plant which bears the distinctly ribbed fruits characteristic of early tomatoes, and this seems to be the earliest printed image of *Lycopersicon esculentum.*

The *Neuw Kreuterbuch* was popular enough so that in 1590 Bassé printed *Eicones plantarum* which is just that, pictures without text, to make a book small enough to carry around. Reusing images in this way was not uncommon; even the fine drawings in Leonhard Fuchs' great folio *De historia stirpium* (1542) were recut in a smaller size and printed in a more easily-used quarto format. In Bassé's little field guide the foreign plants are retained—the maize, several cacti, some squashes, chilies, which had for fifty years been grown as ornamentals, and marigolds, which, too, had become popular ornamentals.

The marigold story is one of name change and mistaken localities which survive in the English common names of "French marigold" for *Tagetes patula* and "African marigold" for *T. erecta,* both species native to Mexico. "Marigold" was the name in English for species of *Calendula,* a European genus of the *Asteraceae,* the daisy family. Calendulas and *Tagetes* marigolds can look much alike, so when *Tagetes patula* was brought into England from France, it simply became the "French" marigold. The other species, *T. erecta,* actually from Mexico, was thought to have been brought to Europe from North Africa in 1535 at the time of the siege of the city of Tunis by Charles V. By the middle of the century *Tagetes erecta* appears in European botanical literature as "flos africanus or aphricanus," the name used by Tabernaemontanus who also gives its German common name of "Thunis blum."

The finely done woodblocks were used again a few years later when they were bought by the London printer, John Norton, who used them in John Gerard's *Herball,* printed in 1597.

Reeds, Karen M. *Botany in Medieval and Renaissance Universities.* New York and London: Garland Publishing, Inc., 1991. See especially Chapter 3, "Botany at the University of Basel in the Sixteenth and Early Seventeenth Century."

29 JOHN GERARD (1545–1612)

The Herball, or General Historie of Plants.

London: E. Bollifant, 1597.
Library of the Gray Herbarium.

Gerard's great work is the best-known herbal in English and is useful and of interest with its first-hand information on plants then grown in England. A barber-surgeon who became master of his company, Gerard early in life became interested in plants and horticulture and was superintendent of the gardens of Lord Burleigh and of a garden owned by the College of Physicians. His own garden, mentioned time upon time in the *Herball,* was famed for its collection of rare and unusual plants.

There is an ancient tale which would make Gerard not the author of this work, but a plagiarist and even thief, and Agnes Arber lays out this story in her *Herbals.* We cannot know the truth in this matter, but Gerard's descriptions of many plants make it clear that he knew them as living things and the information on plants of new world can have come only from Gerard who saw them growing in his garden.

Several American plants are described here. Gerard still includes guaiacum, which seventy years before had been the sovereign remedy for syphilis, but he mentions it only in passing: "whereof our bowles [for bowling] and physicall drinkes are made." His other new world plants are more prosaic species, maize, new world beans, chilies, and the potato.

In his discussion of the potato Gerard created confusion which has plagued the English language ever since. On facing pages are two illustrations. On the left is Chapter 334, headed "Of potatoes;" to the right the heading of Chapter 335 is "Of potatoes of Virginia." Fortunately the illustrations leave no room for doubt. In 1597 the English word "potato," derived from *batata,* referred to the modern sweet potato, *Ipomoea batatas. Ipomoea* belongs to the *Convolvulaceae,* the morning-glory family, and Gerard's picture of the potato shows an *Ipomoea,* with its distinctive undivided leaves.

The plant on the right is just as clearly *Solanum tuberosum,* the modern plain potato. Gerard says of it, "I have received roots hereof from Virginia…which grow and prosper in my garden, as in their

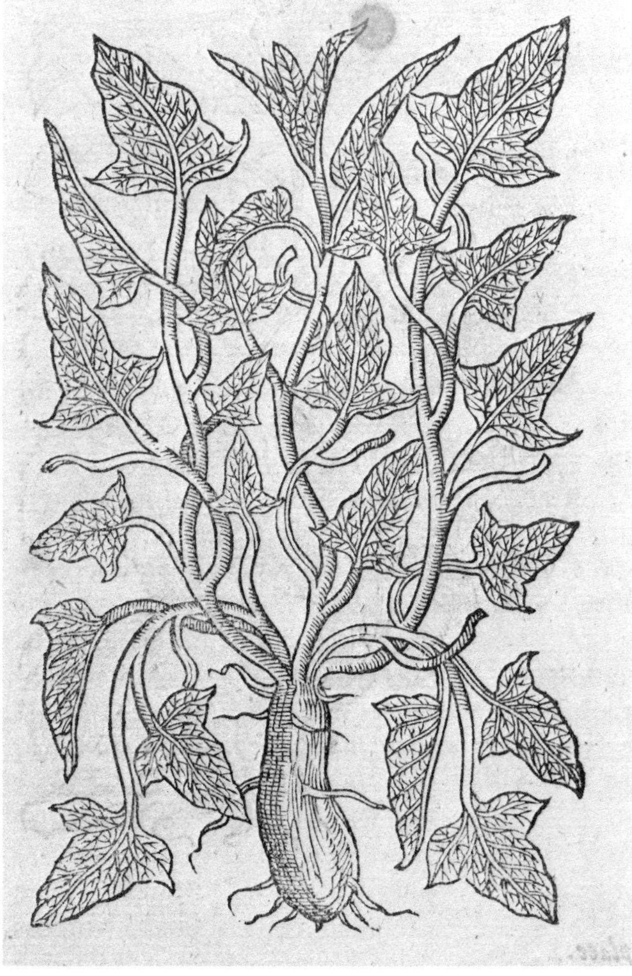

"Potatus or Potatoes," the sweet potato, *Ipomoea batatas.* From J. Gerard,
The Herball. London, 1597. (Catalogue item number 29.)

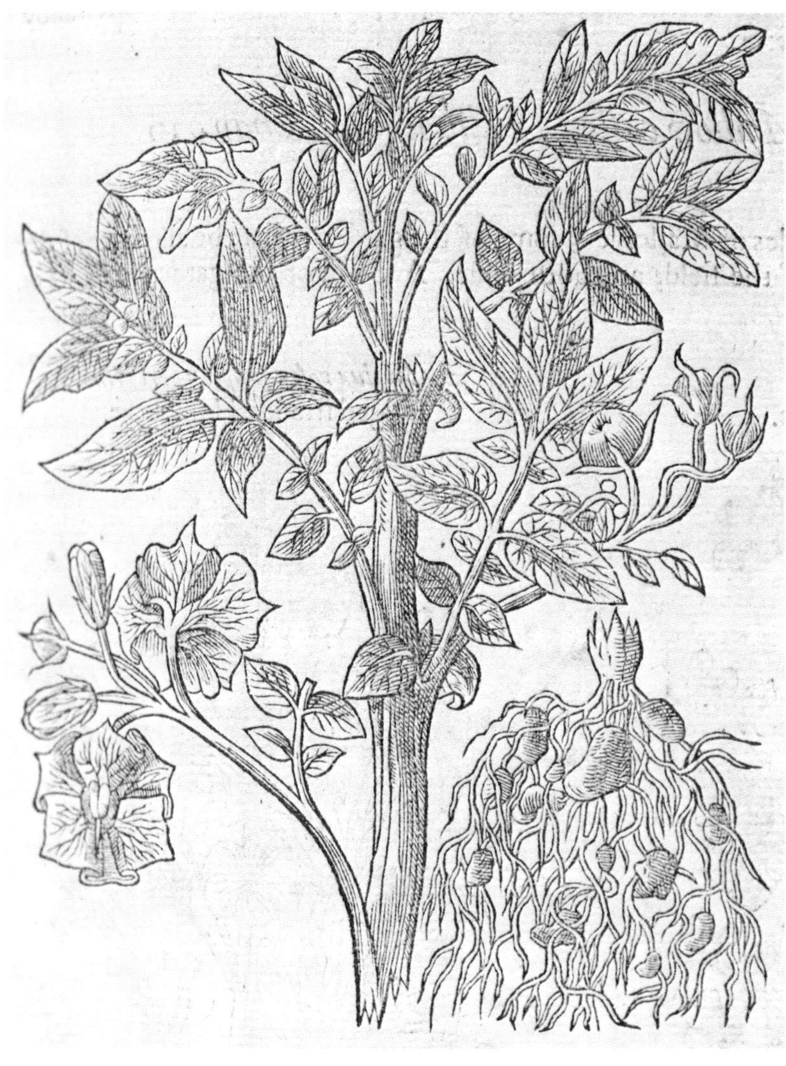

"Potatoes of Virginia," the common potato, *Solanum tuberosum*. From Gerard, *The Herball*. London, 1597. (Catalogue item number 29.)

own native countrie." Then Gerard suggests that since this new plant is like *Ipomoea batatas,* we might "call it in English Potatoes of America, or Virginia." By 1650 the roles were reversed. *Solanum tuberosum* had become the "potato," needing no modifying words, and now *Ipomoea batatas* was the "sweet" potato.

With Gerard began the myth which has potatoes coming from Virginia. His potatoes may have come by way of Virginia, arriving with Sir Francis Drake in 1586. Indeed, Drake called at Roanoke Island in Virginia and brought home the settlers of Walter Raleigh's second expedition, but he was returning to England from Cartagena and any potatoes he brought started their journey in South America.

Arber, Agnes. *Herbals: Their Origin and Evolution.* 3d ed., with an introduction and annotations by William T. Stearn. Cambridge, England: The Cambridge University Press, 1986; esp. pp. 129–135.

Henrey, Blanche. *British Botanical and Horticultural Literature before 1800.* Volume I: "The Sixteenth and Seventeenth Centuries: History and Bibliography." London: Oxford University Press, 1975.

ΠΙΝΑΞ *Theatri Botanici.*

Basel: L. Koenig, 1623.
Library of the Gray Herbarium.

For the modern biologist, the starting date for the nomenclature of most groups of plants is 1 May, 1753, chosen as the common date of publication of the two volumes of Linnaeus' *Species Plantarum.* Of course, Linnaeus did not "invent" botanical nomenclature any more than he "invented" binomials, the two-word combinations of generic name and specific epithet by which we name ourselves, *Homo sapiens,* and the organisms with whom we share the planet. The significance of *Species plantarum* is that the work is a filter through which Linnaeus poured the confused profusion of plant names which had built up by the middle of the eighteenth century—the filtrate being one name for each species.

The profusion of names handled by Linnaeus came from the literature of botany which he knew and used. We too know this literature, for he scrupulously gave the sources for his accepted names and their synonyms. The work most often cited by Linnaeus, appearing under more than 2300 names, is the *Pinax theatri botanici* of Caspar Bauhin.

The *Pinax*—the word here meaning "register" or "list"—is Bauhin's index to the entire world of plants as he knew it—an exhaustive list of names and synonyms, with its own index and a list of the works used by Bauhin who also gave his sources. Called on the title page the result of forty years' work, the *Pinax* is an impressive work and stands as the vantage point from which to look back at the early literature by which knowledge of plants was transmitted. It is especially valuable as a survey point, for Bauhin worked at the end of the sixteenth-century burst of travel and exploration, but before the great influx of extra-European plants which led to proliferation of names during the late seventeenth and early eighteenth centuries.

If we extend the filter analogy, the *Pinax* is the first filter through which plant names passed, and the best guide to the early literature of systematic botany is Bauhin's detailed list of his sources. To compare Bauhin's list (244 works, including multiple editions of some) with that of Linnaeus (252 books and 116 articles in periodicals) is interesting. Bauhin's sources are a diverse lot, much more so than those cited by Linnaeus. Among the works cited by Linnaeus only 36 were printed before 1623, each of these used also by Bauhin; most are the works of well-known writers of herbals.

One problem which arises from these resettings of the informational clock to 00:00 is the loss of knowledge of sources of information. Certainly Bauhin himself ignored many early works which contained information about plants and, in the absence of a summary earlier than 1623, we cannot know with certainty what they were. Research brings to light many of them, but some will be missed. Linnaeus' restart in 1753 dropped out 85% of the works used by Bauhin, and to the modern botanist these works are unknown and effectively lost.

The *Pinax* is a prickly, hedgehog sort of work, bristling with spikes of information and one is loath to see them lost. This exhibition grew, in part, from an attempt to identify the works in Bauhin's detailed, but tersely worded, list of "Nomina authorum quorum opera usi sumus" and to turn that list into a modern "Literature Cited."

Fuchs-Eckert, H.P. "Die Familie Bauhin in Basel." *Bauhinia: Zeitschrift der Basler Botanischen Gesellschaft* 6: 13–48. 1977.

__________. "Caspar Bauhin: Erster ordentlicher Professor der Anatomie und Botanik an der Universität Basel." *Bauhinia* 6: 311–329. 1979; 7:45–62. 1981; 7:135–153. 1982. These papers by Fuchs-Eckert give an exhaustively detailed account of Bauhin's life and career.

"Breve et succinta narratione della navigation…all'Isole di Canada, Hochelaga, Saguenai & altre…" in G.B. Ramusio, *Terzo volume delle navigationi et viaggi.*

Venice: Giunti, 1556.
The Houghton Library. From the bequest of the Hon. William Prescott.

During the sixteenth century there remained the general belief that there existed somewhere a passage through America to the east, and every ocean-going European country looked for it. In 1534, Francis I sent out his expedition, commanded by Cartier, a Breton seaman who had been to Newfoundland and to Brazil. America was not new to France, for French ships had for years gone fishing off Newfoundland and had gone to Brazil for loads of brazilwood.

Cartier made a quick crossing and spent three months exploring the Gulf of St. Lawrence and going into the lower St. Lawrence River. Newfoundland he disliked, calling it, with a frankness unusual in an explorer of the sixteenth century, "surely the land God gave to Cain."

The next spring Cartier returned to Canada with three ships and the charge to explore beyond Newfoundland. This time he sailed into the river, past the Huron town of Stadacone at the site of Québec, and up river to the town of Hochelaga on the Island of Montréal. Cartier's men were warmly welcomed by more than a thousand people who greeted them with gifts of fish and maize bread, throwing so much into the boats that, Cartier said, it seemed to rain bread.

He noted the fields of "corn of the country," maize, which he compares to Brazilian millet—also maize. The Hurons also had in cultivation beans, and fruits which Cartier calls large cucumbers—squashes of some kind. There was, too, a plant which only the men used, carrying the dried leaves and a hollowed stone in a pouch around their necks. The leaves were burned in the hollow of the stone and "they fill their bodies so full of smoke that it streams out of their mouths and nostrils as from a chimney." The French tried tobacco, but thought they had inhaled powdered pepper.

Cartier decided to spend the winter of 1535–36 in Canada and went downriver to the village of Stadacone. Sixteenth-century navigators could determine latitudes quite accurately and he knew that the site was a few degrees south of Paris, but he could have had no idea of a winter spent in a continental climate where his ships would be frozen in for five months.

In December, Cartier heard that 50 people in the Huron village had died of scurvy. By the middle of February fewer than ten of Cartier's 110 men were in good enough health to move around. One day Cartier saw a Huron who had ten days before been very ill, but was now well, healed by the juice of the leaves of some tree. Cartier brought back several branches of a tree called *"ameda"* or *"anedda"* which, when ground up and steeped in water, gave a drink which seemed to cure any disease. After initial reluctance the men found that the stuff worked—even men who had suffered for five or six years from syphilis claimed to be cured. Cartier remarked that they then nearly killed each other to get it and in a week an entire tree had been boiled and drunk.

The tree was, most likely, *Thuja occidentalis,* still called "arbor vitae," the tree-of-life. Cartier took some back to France where fifty years later Charles de l'Escluse saw it in the gardens at Fontainebleau. Cartier had promised for a safe return a pilgrimage to Roc-Amadour and, most likely, he kept that promise. Four centuries later Jacques Rousseau, director of the botanic garden at Montréal, saw Cartier's tree-of-life growing there.

Rousseau, J. "Michel Sarrazin, Jean-François Gaulthier et l'étude prélinnéenne de la flore Canadienne." In *Les botanistes Français en Amérique du Nord avant 1850.* Paris: Centre National de la Récherche Scientifique, 1957; pp. 149–157.

Les voyages du Sieur de Champlain Xaintongeois.

Paris: J. Bergon, 1613.
The Houghton Library.

As early as 1524, Giovanni da Verrazzano, sailing for the French king, had sought without success a northwest passage through America. Ninety years later Samuel Champlain pointed out that France needed a permanent settlement in New France—a place from which exploration for the still unfound passage through the continent could be launched. Earlier attempts had failed, but Pierre du Gua, the Sieur du Monts, a Huguenot nobleman, proposed that he be given a monopoly on the fur trade to support exploration and settlement.

Du Gua's expedition sailed in April, 1604. Among the members was Champlain who had already been to the West Indies and the year before had been to Canada. Champlain, who would be the greatest of early French explorers, spent three years with du Gua. His narrative of these years forms Book I of *Les voyages,* and gives an interesting description of the coasts of Nova Scotia and of New England, especially that of Massachusetts Bay.

Their first winter, 1604–05, was spent on an island in the St. Croix River, which is the modern boundary between the Canadian province of New Brunswick and the state of Maine. It was a hard winter and du Gua's men fell ill with "la malade de terre:" scurvy. Of 79 men, 35 died and more than 20 were near death. The physicians fared as poorly as anyone, and the men who recovered did so only in the spring. Champlain pointed out that it was a change of seasons—actually of diet—rather than medicines. It is surprising that the Indians here had no cure for scurvy. Seventy years earlier, Jacques Cartier's men wintering on the St. Lawrence also had scurvy, but were cured by the Indians' remedy— *Thuja occidentalis,* which so acquired the still-used name "arbor vitae," tree-of-life.

In the spring the survivors sailed southward, and by July they were in Massachusetts waters and in the middle of the month anchored off Duxbury, Massachusetts. Du Gua sent ashore a few men who came back with some small squashes and purslane which were growing in the native people's corn fields. If the plant was indeed purslane, *Portulaca oleracea,* which is generally thought to have been introduced into the new world, its presence suggests an earlier visit by Europeans, weed-carriers par excellence.

On 21 July, they visited a village, probably near Nauset on Cape Cod. As Champlain and the others approached the settlement they walked through a field of maize. "Le bled [bled d'Inde] estoit en fleur de la hauteur de 5. pieds y demy. Nous vismes force febues du Bresil, et force citrouilles de plusieurs grosseurs, bonnes a manger, du petun et des racines, qu'il cultivent, lesquelles ont le gout de artichaut."

The cultivated roots are *Helianthus tuberosus,* the singularly ill-named Jerusalem artichoke, which is not an artichoke and has nothing to do with Jerusalem, for it is not a native of the old world. *Petun,* from the Brazilian name, is tobacco; the word now, as *Petunia,* is the name of a genus of ornamental plants in the *Solanaceae,* the family to which tobacco and tomatoes also belong. And Champlain reports for the first time in northern America the seed-crop trinity of maize, new world beans, and squash which supported life in much of temperate America.

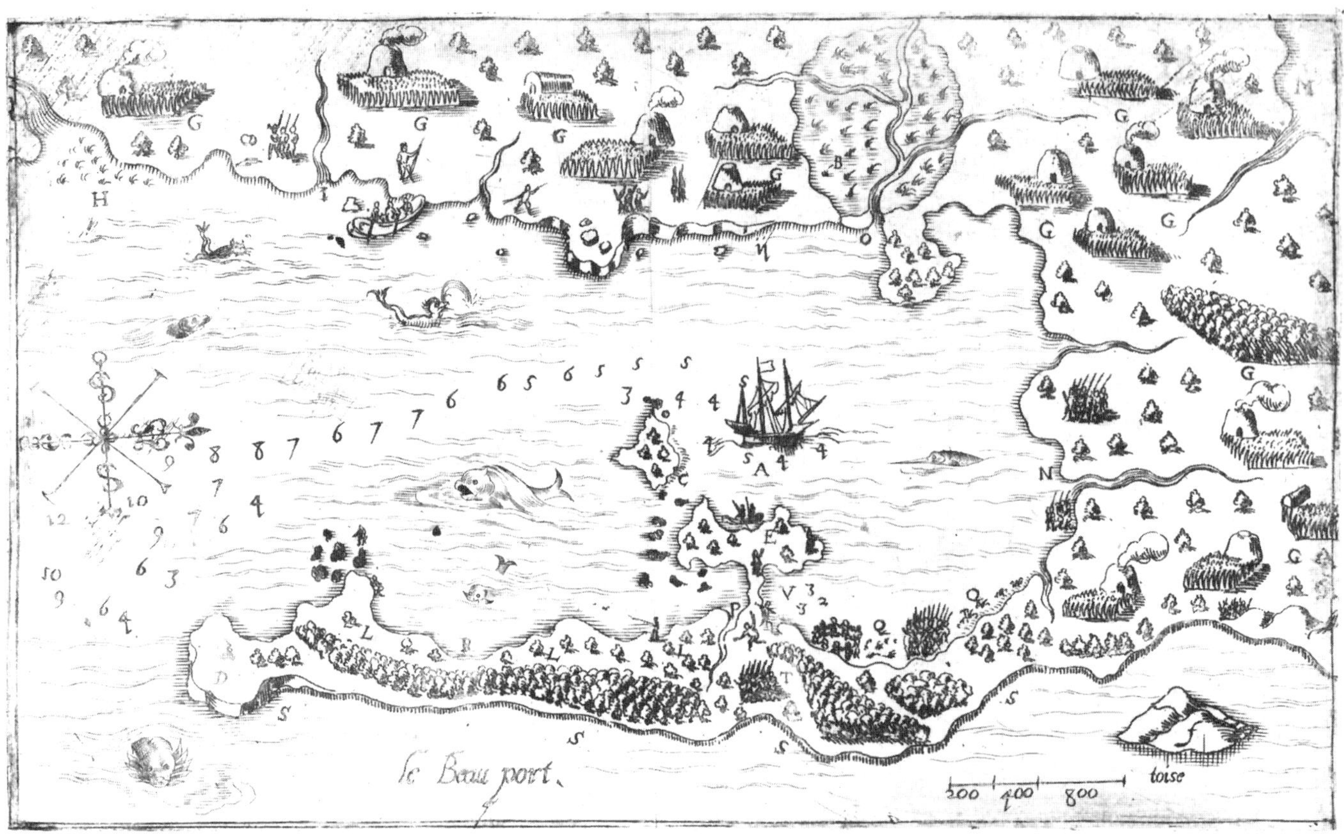

Harbor of "Le Beau Port," Gloucester, Massachusetts; dwellings and crops. From Champlain, *Les voyages.* Paris, 1613. (Catalogue item number 32.)

33   JACQUES CORNUT (1606?–51)

Canadensium plantarum…historia.

Paris: S. le Moyne, 1635.
Library of the Gray Herbarium. Gift of Asa Gray.

During the sixteenth century France made two quite unsuccessful attempts at colonization in the new world. Villegagnon's Huguenot settlement at Rio de Janeiro lasted only a few years (1555–60) and Rene Laudonnière's attempt in 1563–65 to establish a base in Florida was halted by the Spanish. At the beginning of the next century France again looked for an entry point to the new world and found it in the St. Lawrence valley which Jacques Cartier had explored in the 1530's. The first lasting French settlement in America was Québec, founded in 1608 by Samuel de Champlain.

Less than thirty years later, Jacques Cornut published a book which is, in effect, the first work which deals with the plants of a particular part of the new world. Cornut was a Paris physician who never saw the new world, but was able to see American plants newly arrived in France. In France the introduction of foreign plants was an activity more focused than it was in either Spain or England, in that in Paris there was a royal garden which was rapidly becoming a national garden. Both England and Spain had royal gardens but not until much later would there be continuity of effort in introducing and acclimatizing foreign plants. There were also in Paris commercial nurseries, and Cornut acknowledges them as sources of plants.

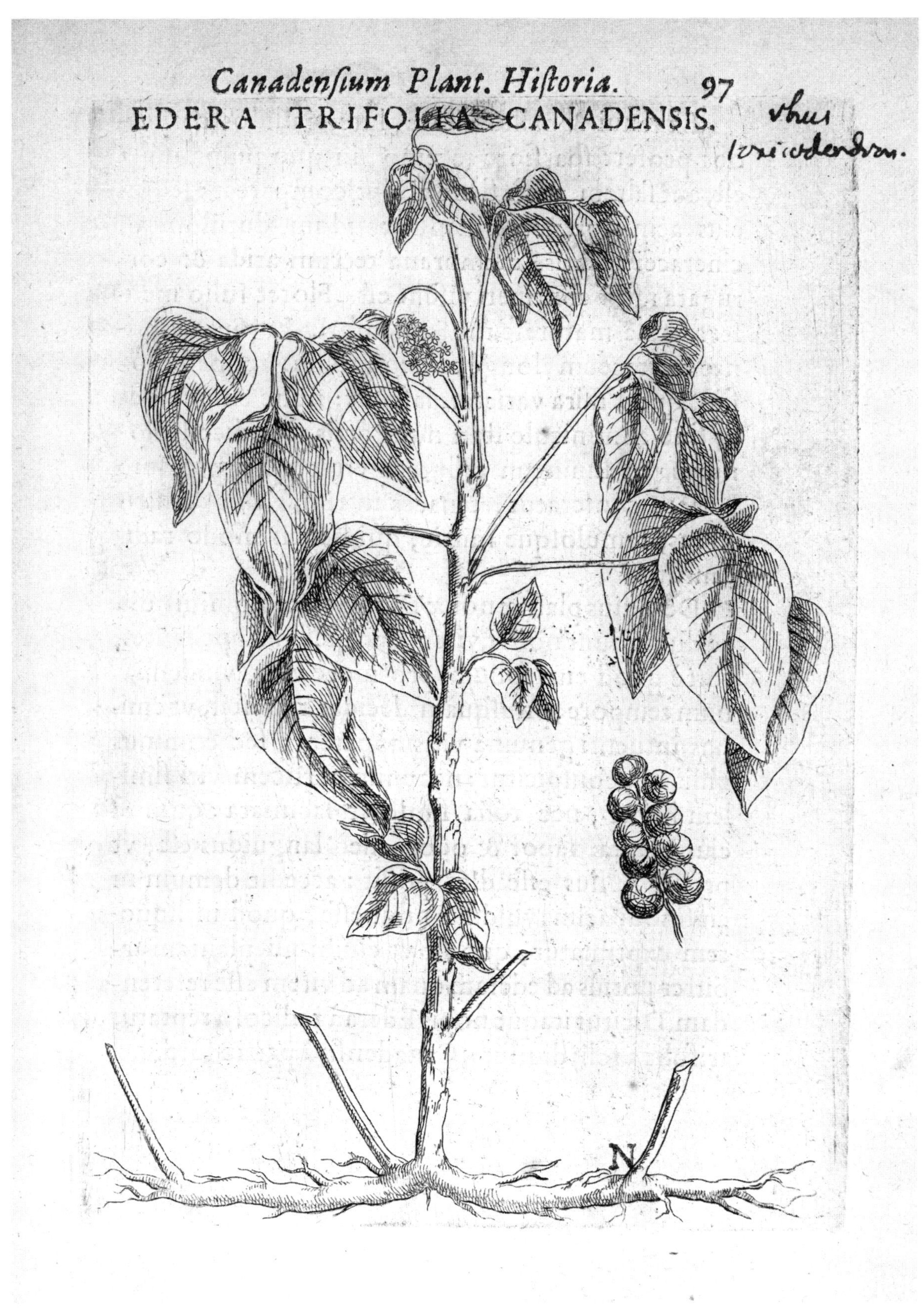

First image of *Rhus radicans*, or poison ivy. From J. Cornut, *Canadensium plantarum...historia*. Paris, 1635. (Catalogue item number 33.)

Smilacina racemosa. From J. Cornut, *Canadensium plantarum…historia.* Paris, 1635. (Catalogue item number 33.)

Cornut described 78 plants each illustrated by an accurate engraving. Of course they are not all Americans, but 44 are, and the engravings are accurate enough that nearly all can be identified. Suddenly the number of new world plants for which the reader could find a good illustration was at least doubled. More than 30 of Cornut's new world plants are here figured for the first time; some are very well-known species to North Americans.

Cornut's "Solanum triphyllum canadense" is *Trillium erectum;* "edera quinquefolia canadensis" is *Parthenocissus quinquefolius,* the commonly grown "Virginia creeper." Many of the plants Cornut saw in the royal garden which was directed by Jean Robin and his son Vespasien. One of these was a handsome tree which Cornut named, as a compliment to the directors, "Acacia americana robini;" it still honors the Robins as *Robinia pseudoacacia,* the North American black-locust. Under the innocuous name "edera trifolia canadensis," three-leaved Canadian ivy, appears for the first time an image of *Rhus radicans,* the all-too-common and familiar poison ivy of North America.

Robbins, William J. "Les botanistes Français et la flore du nordest des États-Unis: J. G. Milbert et Elias Durand." In *Les botanistes Français en Amérique du Nord avant 1850.* Centre National de la Récherche Scientifique. Paris. 1957.

Rousseau, J. "Michel Sarrazin, Jean-François Gaulthier et l'étude prélinnéenne de la flore Canadienne." In *Les botanistes Français en Amérique du Nord avant 1850.* Centre National de la Récherche Scientifique. Paris. 1957.

Stannard, J. "Introduction" to the facsimile reprint of *Jacques Cornut:* Canadensium plantarum…historia. New York and London: Johnson Reprint Corporation, 1966.

34   ROBERT MORISON (1620–83)

Hortus regius Blesensis auctus in praeludia botanica.

London: T. Purfoot, 1669.
Library of the Arnold Arboretum. Gift of Francis Skinner.

Robert Morison fought on the losing side in the English civil wars and, as did so many others, fled to France. In 1648 he received a degree in medicine at Angers and two years later accepted a position as physician to Gaston, Duc d'Orléans, uncle of Louis XIV. Orléans was an avid collector who maintained a botanic garden and a menagerie at Blois, and Morison was involved in curation of the garden. On the death of Orléans in 1660 Morison was able to return to England where he was appointed Royal Physician and Botanist to Charles II; he later became Professor of Botany at Oxford.

Earlier catalogues of Orléans' garden had been printed at Paris in 1653 and in 1655. This third edition, much enlarged, presents to us a great array of plants, impressive even for a royal garden. Morison first lists the plants alphabetically, each with a comment of a few words of reference. He then provides longer descriptions of "species" which were new to Morison or which he thought had not been mentioned by earlier writers.

Here are described, among others, about 20 American plants. One of these is the highly variable *Rhus radicans,* poison ivy, twice described under different names, as "Hedera trifolia Canadensis"—the Canadian ivy with three leaflets—and as "Pistachio virginiana sylvestris trifolia," the wild Virginian pistachio with three leaflets. In fact, pistachios and poison ivy do belong to the same botanical family.

Morison then presents a section in which plants are listed by their uses. Under "evergreens" are "Aloe americana" which is a species of *Agave,* and two kinds of *Opuntia,* the prickly pear. Of all the new world plants which could be grown in European gardens, the strangest, the most alien, and, therefore, desirable, was *Opuntia.*

Under "tuberous" plants are "apios americana," *Glycine apios,* the ground nut of eastern North America, and three kinds of "batattas" [sic]—potatoes. Here Morison's botany is interesting. Two of the potatoes are "Indiae Occidentali…species Convolvuli," which is our sweet potato, and "Virginiana, Solani species," which is *Solanum tuberosum,* our common potato, in its guise as a native of Virginia. The third is a surprise among the potatoes—"Canadensis, Chrysanthemi species"—the Jerusalem artichoke, *Helianthus tuberosus.* It does produce edible underground storage organs and Morison found it logical to place it with potatoes.

Other new world species among "plantae culinariae" are beans, maize, chilies, and nasturtium, now commonly grown as an ornamental. The garden nasturtium has some chemistry in common with *Nasturtium officinale,* the water cress, and they taste much the same. In Morison's time, the garden nasturtium was grown as an ornamental and for the unopened buds which were pickled and used like capers.

35 Arthur Barlowe (fl. 1584)

"The first voyage made to the coastes of America, with two barkes, where in were Captines Master Philip Amadas, and Mastur Arthur Barlow, who discovered part of the Country, now called Virginia, Anno 1584." In Richard Hakluyt, *The Principall Navigations, Voiages and Discoveries of the English Nation…*

London: G. Bishop & R. Newberie, 1589.
The Houghton Library. From the Donation Fund of 1842.

Although there had been an English presence in the new world since the voyages of John Cabot, England was a latecomer to any serious effort to explore and settle in North America, and only in the 1580's were colonists brought to the new world. Sir Humphrey Gilbert tried in 1583, sailing from Plymouth with more than 200 men. They reached Newfoundland, but after many mishaps those who remained returned to England. Gilbert himself was lost at sea off the Azores.

Gilbert's patent from Queen Elizabeth was renewed by his half-brother, Walter Raleigh, who soon sent off two small ships, one of them commanded by Arthur Barlowe. In July, 1584, they landed on the Carolina Banks and then visited Roanoke Island; by the middle of September, they were back in England, bringing them with two Algonquians.

Barlowe's narrative, first published by Richard Hakluyt, is a work of propagandizing which presents their stay on the Banks as an idyll among handsome and goodly people—"for a more kinde and loving people, there can not be found in the world, as farre as we have hitherto had triall." There is some evidence that the truth was a bit less idyllic and that the loving people might have killed as many as thirty of their visitors.

The land is presented as a near Garden of Eden. "The woods are not such as you find in Bohemia, Moscovia, or Hyrcania, barren and fruitless, but the highest and reddest of Cedars of the world, farre bettering the Cedars of the Acores, of the Indies, or of Lybanus, Pynes, Cypres, Sassafras, the Lentisk, or the tree that beareth the Masticke, the tree that beareth the rinde of blacke Sinamon, of which Master Winter brought from the Streights of Magellane, and many others of excellent smell and qualitie."

Barlowe was impressed by the sassafras, which had been known in England since 1577 when John Frampton had translated as *Joyfull Newes from the New Founde Worlde* Nicolás Monardes' book on new medicinal plants from America. Monardes gives a clear description of the distinctive leaf of *Sassafras albidum,* which had already acquired an undeserved fame as a remedy for syphilis; and Barlowe would have easily recognized it.

And the land was amazingly fertile. "Their Countrie corne [maize], which is very white, faire, and well tasted, and groweth three times in five months: in Maye they sowe, in July they reape: in June they sowe, in August they reape; in July they sowe, in September, they reape." Barlowe was an experimentalist—the English planted some of their peas and in ten days the plants were fourteen inches tall.

Indeed, as so described, this Virginia was a place with marvelous natural resources and with a considerable potential for commercial ventures. Preparations for the next voyage were under way even before Barlowe returned, and on the 9th of April, 1585, Raleigh's second expedition sailed to Virginia. That would be a failure and there would be no permanent English settlement in America until Jamestown in 1607.

Parks, George Bruner. *Richard Hakluyt and the English Voyages.* 2d ed. Edited with an introduction by James A. Williamson. New York: Frederick Ungar, 1961.

Quinn, D.B. *The Roanoke Voyages.* Hakluyt Society Publications, 2d series, Vol. CIV. London, 1955.

36   THOMAS HARIOT (1560–1621)

A Briefe and True Report of the New Found Land of Virginia.

Frankfurt am Main: J. Wechsel, 1590.
The Houghton Library. From the bequest of the Hon. William Prescott.

In April, 1585, the second of the expeditions sponsored by Sir Walter Raleigh sailed with several ships and some 500 men, 108 of whom were to be colonists. After some adventures in the West Indies, they reached "Virginia" late in June. In the party were an artist, John White, who earlier had been in the Arctic with Martin Frobisher; and a young man, Thomas Hariot, whose assignment from Raleigh was to observe and to map as much as possible and to prepare a report. The settlement was made on Roanoke Island but relations between the settlers and the native people deteriorated by the following summer to the point of an armed assault by the English.

In June, 1586, a fleet under Sir Francis Drake appeared, returning to England from piracy in the West Indies. At the entreaty of the colony's governor he took aboard the remaining colonists and returned them to England.

Hariot's report was printed in 1588 as *A Briefe and True Report,* known now by only a few copies. The *Briefe Report* was soon reprinted by Théodore de Bry of Frankfurt, the first in the long series of narratives of exploration and travel known to us as de Bry's "America" or "les grands voyages."

The report was written, in part, as propaganda to encourage settlement in Raleigh's new colony, but it is also a superb description of the land and of its people—and it contains the first overview of the plants and animals of any part of temperate America. The propagandizing function of the report becomes apparent immediately after the introduction—"The first part of marchantible commodities."

The first two of these listed, "Silke of grass" and "Worme Silke" are hardly "marchantible." The silk of grass is, most likely, from a species of *Yucca,* and "worme silke" may be the communal webs of the eastern tent caterpillar, *Malacosoma americanum.* But Hariot also reports sassafras which had been first made known in Europe by Nicolás Monardes (1571), and he refers to John Frampton's translation, *Joyfull Newes out of the Newe Founde Worlde,* printed in 1577. Sassafras had already acquired a reputation, quite undeserved, as a cure for syphilis.

Hariot reported carefully the foods of the local people. First is maize and Hariot gives a good description, noting what most earlier writers had remarked upon—"It is a graine of marveillous great increase; of a thousand, fifteen hundred and some two thousand fold." Cultivated, too, was *uppowoc,* "of so precious estimation amongst them, that they thinke their gods are marvelously delighted therwith." This is tobacco and Hariot describes the first use of clay pipes as a smoking aid. Twenty years later the colony at Jamestown would be established and clay pipes of differing styles would become so commonly used that they are now a useful aid in dating seventeenth-century sites in North America.

Bucher, Bernadette. *Icon and Conquest: A Structural Analysis of the Illustrations in de Bry's Great Voyages.* Translated by Basia Miller Gulati. Chicago: University of Chicago Press, 1981.

Shirley, John W. *Thomas Harriot: A Biography.* Oxford: Clarendon Press, 1983.

37   JOHN BRERETON (fl. 1602)

A Briefe and True Relation of the Discoveries of the North Part of Virginia.

London: G. Bishop, 1602.
The Houghton Library. From the bequest of the Hon. William Prescott.

At the end of the sixteenth century England's chief promoter of activity in North America was Sir Walter Raleigh. His attempted colony in Roanoke Island in Virginia had failed and the settlers left behind had simply disappeared; his interests had turned, in part, to developing a trade in sassafras which was still a valuable commodity in England. To this end, Raleigh had received from the Queen a patent which gave him a monopoly on the importation of sassafras into the country.

Late in the summer of 1602 the price of sassafras fell and Raleigh found that a quantity had been unloaded at Southampton and offered for sale in London. The vendor was Bartholomew Gilbert who had just returned from America.

The first account of this "unauthorized" voyage, led by Bartholomew Gosnold, was Brereton's work. Brereton, a clergyman, was one of the party, but in what capacity we do not know. Gosnold sailed in March, 1602, "holding a course for the North part of Virginia"—actually New England. Landfall was made on the coast of Maine, but Gosnold sailed southward and the very next afternoon found them in Massachusetts Bay. After sailing around the tip of Cape Cod (so named by Gosnold, for in a few hours fishing "we had pestered our ship so with Cod fish, that we threw numbers of them overboard againe"), serious activity began on the south side of the Cape.

Most of their exploration was on the island of Martha's Vineyard, generally thought to have been named by Gosnold for his daughter, and on Cuttyhunk, southernmost of the Elizabeth Islands which run southward from the western end of Cape Cod.

From the Vineyard, Brereton reported beeches and cedars and an abundance of berries, "and such an incredible store of Vines, as well in the woodie part of the Island, where they run upon every tree, as in the outward parts, that we could not go for treading upon them." On Cuttyhunk, which Gosnold named Elizabeth's Island, they finally found sassafras, "great plentie all the Island over, a tree of high price and profit." There they built a house, cut sassafras, and traded for furs with the people of the mainland.

There had been some thought of leaving a small party on Cuttyhunk, but in the end, the ship loaded with sassafras, cedar, and furs, all hands returned to England. Raleigh's anger about the load of sassafras was assuaged, with a bit of legal action involved as well. By the time of printing an agreement had been reached that the title page would show that the voyage was made with Raleigh's permission.

Fogg, John M., Jr. "The Flora of the Elizabeth Islands, Massachusetts." *Rhodora* 32: 119–132, 147–161, 167–180, 208–221, 226–258, 263–281. 1930. Fogg notes that although Brereton speaks of the islands as being densely wooded, several are now treeless.

38   JOHN SMITH (1580–1631)

A Description of New England.

London: H. Lownes, for R. Clarke, 1616.
The Houghton Library. Gift of Samuel A. Eliot.

John Smith is, to most North American readers, firmly associated with the English settlement made in 1607 at Jamestown, Virginia, and regarded as the man who brought the colony through its uncertain first year. When Smith reached Jamestown he had already spent the most adventuresome part of his life. In 1600 Smith had gone to join Austrian forces fighting the Turks; he was captured in Transylvania and taken as far as the Black Sea. Here he escaped, returned via Russia and Poland to Transylvania, and by the winter of 1604–05 was back in London.

Smith spent two years in Virginia and returned again to England, in part because of injury and, in part, because he seems to have been a prickly and difficult man who had exhausted his welcome. His next foreign travel was in 1614 to North America where his party explored the coast of Maine and of Massachusetts Bay, and Smith, with royal permission, named the country "New England."

As was Thomas Hariot's *Briefe Relation* a mixture of propaganda and description, so is *A Description of New England*. Smith saw himself as both explorer and promoter of English settlement in North America and this little book is on the side of promotion rather than extensive description. His expedition arrived in April, 1614, at Monhegan Island off the coast of Maine to take whales and to look for gold and copper; if these failed it was to be fish and furs.

In fact, gold, copper, and whales did fail, it was too late for furs, and the reality of fishing seemed not to interest the adventuresome Smith, so they sailed south along the coast of southern Maine and into Massachusetts Bay inside the curved arm of Cape Cod. Smith noted the country, remarking often upon Indian maize fields; of these he seemed acutely aware, perhaps because his ability to beg corn from the Indians at Jamestown had helped to save the colony. Smith is, in general, matter-of-fact about the coast and its products. He does note that "first, the ground is so fertill, that questionless it is capable of producing any Grain, Fruits, or Seed you will sow or plant," and remarks that in May he planted on a rocky island a garden which in June and July provided salad stuff.

Smith comments upon several kinds of trees, including "saxifrage," a lapsus of some kind for sassafras which still was a product valuable enough to attract the attention of an explorer. A note of eloquence shows up on page 26 when he speaks of "the Countrie of the Massachusets, which is the Paradise of all those parts: for heere are many Iles all planted with corne, groves, mulberries, salvage gardens, and good harbors."

In his role as propagandist Smith notes how successful Spain and Portugal have been in both the East and West Indies and suggests that England might seek to do as well; "And seeing, for all they have, they cease not still to search for that they have not, and know not; it is strange we should be so dull, as not maintaine that which we have, and pursue that wee know." Six years later the first permanent settlement in New England was made at Plymouth, Massachusetts.

39   EMANUEL SWEERTS (1552–1612)

Florilegium.

Frankfurt am Main: Part 1, Antony Kempner; Part 2, Erasmus Kempffer, 1612.
Library of the Gray Herbarium.

Florilegia, meaning gatherings of flowers, became popular at the end of the sixteenth century as sets of paintings, usually watercolors, and as printed books. As a book, the florilegium is, most often, a collection of engraved plates, usually with very little text and often with only the names of the plants. The first florilegia of this sort were intended as sources of designs, which one might use as embroidery patterns or as a background for one's own attempts at flower painting.

In 1612 two florilegia were printed, one the work of Jean-Théodore de Bry, the other by Emanuel Sweerts. Sweerts himself was something of an artist, but his chief activity was the selling of novelties of all kinds, exotic plants, and things such as crystals, armadillo shells, Tupinambá Indian artifacts from Brazil—whatever might attract the attention of a well-to-do person who kept a cabinet of oddities.

Sweerts spent most of his life in Amsterdam but he had close ties with central Europe, for he had been supplier of plants to the eccentric Emperor Rudolf II of Austria. His florilegium in its first two editions (1612 and 1614) is avowedly a nursery catalogue, although later editions (it was reprinted as late as 1655) were produced as mere collections of plates.

His natural history business was based in Amsterdam, but each year Sweerts went to the Frankfurt Book Fair; his advertisements, in Latin, French, Dutch, and German, suggest that he who might wish to buy the book or the plants contained in it should go at fair time to the author's shop opposite the town hall in Frankfurt.

Europe in 1612 was at the start of the great influx of foreign plants which would continue for the next two hundred years, and rare and interesting plants from the Ottoman Empire were beginning to appear in European gardens. During his years in Vienna, the Flemish botanist Charles de l'Escluse had introduced into horticulture many of the bulbous plants now commonly grown: tulips, narcissus, crocuses, and that handsome tree, *Aesculus hippocastaneus,* the horse-chestnut. The Dutch had established a base at Cape Town as well and spectacular plants from South Africa were being brought back to Europe.

The first part of Sweerts' catalogue, 67 plates, is devoted to "bulbs" and he presents an interesting array with many kinds of *Narcissus* and *Iris.* He has also some of the virus-infected Rembrandt tulips with their showy stripes. One bulb, perhaps a species of *Zephyranthes,* he calls "narcissus of Virginia" received from Peter Garet of London.

The second part, 43 plates of plants "with roots" (as opposed to "with bulbs"), includes more than a dozen American plants. Assuming that the plants of the catalogue were in fact available, one could get two kinds of cacti, tobacco, tomatoes, chilies, sunflowers, nasturtiums, French and African marigolds, marvels-of-Peru, pineapples—although one wonders if Sweerts could really supply these—and sweet and white potatoes. All of these were by now familiar to Europeans, even if only by reading about them. Maize is not listed, but it had, perhaps, outgrown its period of favor as a novelty for the garden.

40   PIERRE VALLET (ca. 1575–?)

Le jardin du roy tres Chrestien Loys XIII…

Paris: 1623.
Library of the Arnold Arboretum. Gift of Sarah C. Sears.

Toward the end of the sixteenth century there began to appear in botanical books images printed from metal plates, some of them engravings, some etchings, some a combination of techniques. The first

"Nasturtium alterum," a species of *Tropaeolum*, the garden nasturtium, native to the northern Andes. From E. Sweerts, *Florilegium.* Frankfurt a.M., 1612. (Catalogue item number 39.)

botanical books in which etchings were used are Joachim Camerarius' *Symbola et Emblemata* (1590–1604) and the *Phytobasanos* of Fabio Colonna (1592).

During the next twenty years there developed a genre of book, the florilegium. The word is coined from "flores," flowers, and "legere," to collect, and refers, in this context, to a collection of plates, accompanied by comparatively little text, sometimes only the names of the plants. During the seventeenth century these collections of plates, usually beautifully engraved, became very popular and "flower books," descendants of the florilegia, are still produced today.

One of the earliest and least known florilegia is *Le jardin du roy tres Chrestien Henry IV…dédie a la*

The potato, *Solanum tuberosum*. From Sweerts, *Florilegium*. Frankfurt a.M., 1612. (Catalogue item number 39.)

royne of Pierre Vallet (1608). The book on exhibit here is the second edition of 1623, the change in title reflecting the fact that Louis XIII was now King. The two editions have many plates in common, but the edition of 1623 has added several which show new world plants.

Vallet is someone of whom we know little, other than that he was a talented artist who could work with accuracy in several media. The frontispiece, probably engraved by Vallet himself, shows the artist and in each corner of the print, a set of tools of one of his trades—the draughtsman's compass and square, the engraver's burins, the painter's easel and palette, and the embroiderer's needles and bobbins. Vallet signed himself with pride as "brodeur ordinaire du roy."

Le jardin is dedicated to Marie de Médicis, as Queen in the first edition; in the second, as Queen Mother. Among her interests was needlework and the plates are intended as embroidery patterns, or as a basis for other artistic work. Vallet presents an advertisement to those "qui voudront peindre ou anluminer, ou broder et faire tapisserie sur le present livre." At the end he provides sixteen pages of advice on the correct colors to be used; for "Trachelium Americana," which is the new world cardinal flower *Lobelia cardinalis:* "elle est d'un rouge vif comme escarlatte." Other new world plants presented are *Lilium canadense* and the passion flower. They are not in the first edition; their presence here suggests that Vallet met them in the gardens of the Louvre where the King's gardener, Vespasien Robin, introduced many plants from abroad.

Blunt, Wilfrid. *The Art of Botanical Illustration.* London: Collins, 1950.

Nissen, Claus. *Die botanische Buchillustration.* 2d ed. Stuttgart: A. Hiersemann, 1963.

41   JOACHIM CAMERARIUS (1534–98)

Symbolorum et emblematum ex re herbaria…collecta.

Nuremberg: J. Hoffmann & H. Camoxius, 1590.
The Houghton Library. Gift of Philip Hofer.

Camerarius, son of the better-known humanist and philologist of the same name, spent his life in Nuremberg in the practice of medicine and in translating and editing earlier texts. In 1588 he published a botanical/medical work, *Hortus medicus et philosophicus,* but Camerarius is not a major figure among herbal writers.

Although it seems not to be mentioned in histories of botany, Camerarius published four emblem books, all printed at Nuremburg. The first "century" of emblems is drawn from plants (1590), the second century of emblems is quadrupeds (1593); the third century is flying creatures (1596), and the fourth (1604) is fish and reptiles. Emblem books were popular during the sixteenth and seventeenth centuries and Camerarius' "centuries" were several times reprinted.

The first true emblem book was Alciato's *Emblematum liber* (1531) in which pictures are combined with earlier written epigrams. By the middle of the sixteenth century emblem books had proliferated, providing combinations of *pictura,* a symbolic picture; *inscriptio,* a short motto; and *subscriptio,* a short passage of verse or prose. The combination of visual and verbal elements expressed some moral truth which the memory could readily seize and hold fast.

Camerarius' emblems are drawn, for the most part, from the world of real natural history; there are no sciadopods, no satyrs, no winged tongues. A few of his emblems are based upon organisms of the new world. In the first century we find the sunflower, turned towards the sun, and in the second century (quadrupeds) there is the armadillo, clad in its loricate armor, which delighted and impressed the Europeans who first met the animal.

Camerarius includes a strong component of up-to-date natural history. For each emblem there is a commentary which includes the sources for his descriptions and interpretation. Camerarius uses the usual array of Renaissance sources, from Solomon to Isidore of Seville, to Alciato, but, as well, contemporary writers on the new world.

It is interesting to find new world organisms used in emblems. Living things, especially animals, of the old world, carried with them a huge baggage of associations, legends, etymological facts and fancies, and facts and non-facts of natural history, but American plants and animals came unencumbered by those associations which made a "good" emblem. They were new and stood bare to European eyes and minds, for there were not present the similitudes and resemblances which were so important a part of sixteenth-century natural history. Some were, however, too good to ignore, and the sunflower, the giant plant with its face turned towards heaven, was one.

As Calderón wrote of the sunflower, "Aquel girasol, que está/ viendo cara á cara al sol/ Tras cuyo hermoso arrebol/ siempre moviendose va." Its supposed following of the sun made it the perfect image to remind the viewer of man's duty always to turn toward the source of heavenly grace.

Ashworth, William B., Jr. "Natural History and the Emblematic World View." In David C. Lindberg & Robert S. Westman (eds.), *Reappraisals of the Scientific Revolution*. Cambridge, England: Cambridge University Press, 1990.

Harms, Wolfgang. "On Natural History and Emblematics in the 16th Century." In A. Ellenius (ed.), *The Natural Sciences and the Arts*. Stockholm: Almqvist & Wiksell International, 1985.

42 WILLIAM SHAKESPEARE (1564–1616)

The Merry Wives of Windsor.

London: T. H[arper] for R. Meighen, 1630.
The Houghton Library. Part of a gift made by members of the family of William A. White in 1928–29.

One of Shakespeare's great comic characters is the witty and free-spirited fat knight, Sir John Falstaff. He appears in the two parts of *Henry IV* as a father-substitute, familiar with taverns and revelry, for young Prince Hal who will become Henry V, but at the end of *Henry IV, Part 2,* Falstaff is rebuffed by the newly crowned prince, and in *Henry V,* we hear of his death.

Falstaff appears, as well, in the farcical comedy, *The Merry Wives of Windsor,* written, so the story goes, at the request of Queen Elizabeth who wished to see Sir John in love. The main plot concerns the consequences of Falstaff's simultaneous sending of love letters to Mistress Ford and Mistress Page, married ladies of Windsor, who decide that he must be punished for this presumption. Eventually Falstaff is lured by the merry wives, their husbands and friends to Windsor Forest for a supposed midnight assignation with Mistress Ford, which will turn out to be his come-uppance at their hands.

As Mistress Ford approaches, Falstaff has these lines: "Let the sky rain potatoes, let it thunder to the tune of 'Greensleeves,' hail kissing-comfits, and snow eryngoes; let there come a tempest of provocation, I will shelter me here." But why should Falstaff call for a rain of potatoes and a snow of eryngoes during this midnight tryst?

The Merry Wives is thought to have been written during the early months of 1597 and it is likely that the potatoes and eryngoes came to Shakespeare from John Gerard's *Herball*, printed in the same year. It is clear from the *Herball* that the word "potato" meant our modern sweet potato, *Ipomoea batatas,* which had a quite undeserved reputation as an aphrodisiac.

Gerard speaks in glowing terms of "potatoes." "These rootes may serve as a ground or foundation, wherein the cunning confectioner or Sugar baker may worke and frame many comfortable delicate conserves, and restorative sweetmeates. They are used to be eaten rosted in the ashes; some when they be so rosted, infuse them, and sop them in wine: and others, to give them the greater grace in eating, do boile them with prunes. and so eat them. And likewise others dresse them (being first rosted) with oile, vineger and salt, every man according to his own taste and liking: notwithstanding howsoever they be dressed, they comfort, nourish, and strengthen the bodie, procure bodily lust, and that with greedinesse."

The eryngoes are also reputed aphrodisiacs, but good native English ones: the candied roots of the sea-holly, *Eryngium maritimum.* Gerard recommends them too; "it is also good for other sorts of people that have no delight or appetite to venery, nourishing and restoring the aged, and amending the defects of nature in the younger."

Craik, T.W. (ed.) *The Merry Wives of Windsor.* Oxford: Clarendon Press, 1989.

Salaman, Redcliffe N. *The History and Social Influence of the Potato.* Cambridge, England: Cambridge University Press, 1985.

43   Jan de Laet (1593–1649)

Novus Orbis seu descriptionis Indiae Occidentalis Libri XVIII.

Leyden: The Elzevirs, 1633.
The Houghton Library. Gift of the Society for Propagating the Gospel in New England.

Less well-known and less successful than its older East Indian counterpart, the Dutch West India Company was established in 1621 with the goal of regulating for the national good the illicit trade which the Dutch carried on with Spanish and Portuguese possessions in the new world and in west Africa. Spain and Portugal were then politically united and a blow struck at Portuguese colonies was a blow struck at Spain, with whom the Dutch Republic had been at war since its proclamation of independence in 1579.

The West India Company was given a monopoly of trade with America and the western coast of Africa, with much the same privileges as the East India Company—the right to negotiate treaties, to make war (and peace) with native rulers, and to appoint governors. The company was controlled by a board of directors, the Heeren XIX, one of whom was Jan de Laet, a well-known and respected scholar.

There was no modern summary of the Americas to serve as a baseline for the Company and de Laet set out to produce one. This he did with Dutch thoroughness and produced a marvelous book which is laid out—and reads—as if it had been produced a hundred years later. It was first printed in 1625 as *Nieuwe Wereldt* and enlarged in 1630 as *Beschrijvinghe van West-Indien,* from which this book was translated. The new world is parcelled out into eighteen books, each book concerned with a geographical region, and broken into chapters which note both the major and minor voyages there, the country, the people, and the plants and animals. In addition there are fourteen large maps which have no marginal vignettes, no monsters filling the expanses of the ocean, just straightforward geographical information.

Two of the eighteen books are on Brazil and it is here that de Laet gives the reader the most natural history, much of it drawn from André Thevet's *Les singularitez de la France Antarctique* (1557) and Jean de Léry's *Histoire d'un voyage faict en la terre du Brésil* (1578). The emphasis lies on Brazil, for the Company had already targeted it as a vulnerable part of the Spanish-Portuguese empire. After a brief occupation of Bahia in 1624–25, the Company turned its attention to the sugar-rich region around Pernambuco and fought its way into Recife in 1630.

De Laet provides detailed descriptions of several American plants, notably maize, cashews and cacao, and of the old world plant, sugar cane, which had become the first American plantation crop. To compare de Laet's discussions of new world plants with the treatment in, for example, Dalechamps' *Historia* of fifty years before is to compare a world of plants which are, for the most part, poorly known oddities, things at which the botanist might wonder, with a world in which the plants are about to become monocultured commodities, colonial raw materials to feed the mercantile interests of a colonizing mother country. The purpose of the West India Company is neatly summarized in the vignette at the bottom of the titlepage. Three Brazilian Indians, accompanied by an armadillo, are offering the riches of the new world to a female figure, labelled "fœd. belg.," who has at her side a sword and musket.

44   [CÉSAR *or* CHARLES] DE ROCHEFORT (fl. 1658)

Histoire naturelle et morale de les Iles Antilles.

Rotterdam: A. Leers, 1665.
The Houghton Library. Gift of Frances Hofer.

On the title page of this book there is no author's name, but the dedication is signed by "M. de Rochefort." However, the book is listed in Barbier's *Dictionnaire des ouvrages anonymes*, and Barbier says that "de Rochefort" is a pseudonym of one Louis de Poincy. Whoever the author was, he had personal experience of the French West Indies and a lively interest in what he had seen there.

The first twenty-four chapters deal with natural history and the author—let's call him "de Rochefort"—works his way through geography, botany, and zoology. Six chapters are concerned with plants, both native and introduced, with brief mentions of sugar cane, ginger, and indigo, all three old world species which were grown as crops. De Rochefort barely mentions maize, except to mention that it is cultivated; most likely, maize was well enough known to be of no particular interest.

There is an interesting note on a species introduced to Europe from North America. De Rochefort has a chapter on "patates" which, as is clear from the illustration, are sweet potatoes. They are presented as if unfamiliar to the reader and the author describes the roots by comparing them with those "trufes des jardins" which are called "toupinambous" or "artichauts d'Inde," but states that the "patates" are much better and are better for you. He digresses to point out that toupinambous are today very common and "fort vils & fort méprisez," even though they were once rare delicacies. The despised toupinambous are Jerusalem artichokes, *Helianthus tuberosus,* related to the sunflower, but grown for their tuberous roots and the name for this species in both French and English shows how misleading common names can be. "Jerusalem" has nothing to do with a place, but is a garbled "girasole." "Toupinambou" comes from Tupinambá, the tribe of Brazilian Indians best known to Europeans, and is a case of mistaken locality. The newly introduced plant was thought to have come from Brazil, rather than from Canada.

The second twenty-four chapters are concerned with people and with the economy. In a chapter headed "Du trafic & des occupations des habitans etrangers du pais: & premierement de la culture & de la preparation de tabac," de Rochefort explains that the islands produce five chief items of exchange; tobacco, sugar, ginger, indigo, and cotton. The first of these to be cultivated was tobacco, native to the Americas, but when supply exceeded demand, planters turned to three introduced species, sugar cane, ginger, and indigo. "C'est une merveille de voir avec quel succés, toutes ces marchandises croissent en la plu-part de ces Iles." Although these species were mentioned in the chapters on natural history, they appear here in much more detail as the commodities they had become.

Barbier, Antoine Alexandre. *Dictionnaire des ouvrages anonymes.* Paris: G.P. Maisonneuve & Larose, 1964.

ABOVE: Grating of cassava, *Manihot esculenta.* BELOW: The sensitive plant, *Mimosa pudica.* From C. de Rochefort, *Histoire naturelle et morale.* Rotterdam, 1665. (Catalogue item number 44.)

ABOVE: *"Urucú," Bixa orellana,* a widely-used source of body paint. BELOW: Coconut, *Cocos nucifera,* early introduced into America. From de Rochefort, *Histoire naturelle et morale.* Rotterdam, 1665. (Catalogue item number 44.)

Gathering of papayas, *Carica papaya*. From de Rochefort, *Histoire naturelle et morale*. Rotterdam, 1665. (Catalogue item number 44.)

45   PHILIPPE-SYLVESTRE DUFOUR (1622–87)

Traitez nouveaux & curieux du café, du thé, et du chocolate.

The Hague: A. Moetjens, 1685.
The Economic Botany Library of Oakes Ames.

About 1660 there began to appear works which dealt with coffee, tea, or chocolate, often with all three of these alkaloid-containing drinks. Coffee and tea, both products of the old world, became familiar to European consumers during the 1640's and 1650's.

Chocolate, an American native, had been used in Spain since the 1540's, but achieved wider-spread popularity only after the middle of the seventeenth century. One hundred years later, Linnaeus gave to the tree whose seeds give us chocolate, the marvelous name *Theobroma cacao*: "cacao, food of the gods."

Philippe Dufour was a dealer in medicinal products who had a broad interest in literature and the arts. His little book on the three alkaloidal drinks was first printed in Lyon in 1671. It is not original, for each of the three parts is modified from an earlier published work, that on chocolate from a treatise by Antonio Colmenero, a Spanish physician. Dufour says plainly that he would have liked to see these plants in their native places, but the way there is too long and too dangerous, so he will just refer to what others have to say. The book was widely popular and was even translated into Latin (The Hague, 1693).

Dufour gives a recipe for the near-solid paste which was mixed with hot water to make the drink: cacao beans, sugar, cinnamon, chilies, cloves, vanilla or anise, and *achiote* (the fruit of a West Indian tree, *Bixa orellana)*.

The paste was molded into tablets which were tightly wrapped in paper and could then be kept for weeks. Dufour remarks that sugar was used both to improve the flavor and as a preservative. Other things, such as orange flowers and ground almonds, could be added to the mixture. Quoting from Colmenero on those who would add maize flour to the mix, as did the Aztecs, Dufour adds "très mal," for this mixture engenders "l'humeur mélancholique."

It is worthy of note that the great increase in use of the alkaloid-containing drinks, which are bitter, or in the case of tea, both bitter and astringent, comes soon after the price of sugar in Europe started to fall. Sugar cane was grown very early in Hispaniola, and in Brazil from about 1530, but in Europe the product was expensive. Sugar was used in medicine, or as an expensive sweetener, or in the construction of "subtilitées," giant confections which graced the tables of those who could afford them.

England started sugar growing in Barbados and in Jamaica during the 1640's; the sugar supply increased and consumption rose. Consumption of sugar in England increased from 300 barrels in 1660 to 50,000 in 1700, and the pattern of use of hot, sweetened, alkaloid-containing drinks was established. They are both stimulants and providers of calories and their popularity continues.

46 MICHAEL BOYM (1612–59)

Flora Sinensis.

Vienna: M. Richter, 1656.
Library of the Arnold Arboretum.

Michael Boym's *Flora Sinensis* is the first western work on Chinese plants and, according to Alphonse de Candolle in *La phytographie* (1880), is the first work to use "flora" in the sense of an inventory or summary of the plants of a particular region.

Boym, son of the physician to Sigismund III of Poland, went as a Jesuit to China in 1643, arriving at a time of turmoil and rebellion. In 1644 Beijing fell to rebel forces and soon was under control of a Manchu emperor. Some of the ousted Ming family—but not the emperor—and many members of the household were Christians, converted by Jesuits who had been active in China since the days of Matteo Ricci. In 1650 Boym, who had been a missionary on the island of Hainan, became attached to the court of the last Ming ruler, the Yung-Li emperor, who still controlled part of southern China.

The dowager empress decided to call upon the Pope and the Superior General of the Society of Jesus for their prayers, and for more missionaries. To this end, in 1652 she sent letters to Europe, the letters carried by Boym who had near-ambassadorial status. After long delay, the letters were answered and the answers carried back, but by the time that Boym returned in 1658, most of the Ming were dead. Boym himself died of ill-health and grief in southern China in 1659.

During his years in Europe Boym had several works published. This one is among the rarest of botanical works for only ten or eleven copies are known. There are sixteen engravings of plants and four

Cashew, *Anacardium occidentale,* showing the enlarged fruit stalk ("cashew apple") and the terminal fruit. From M. Boym, *Flora sinensis.* Vienna, 1656. (Catalogue item number 46.)

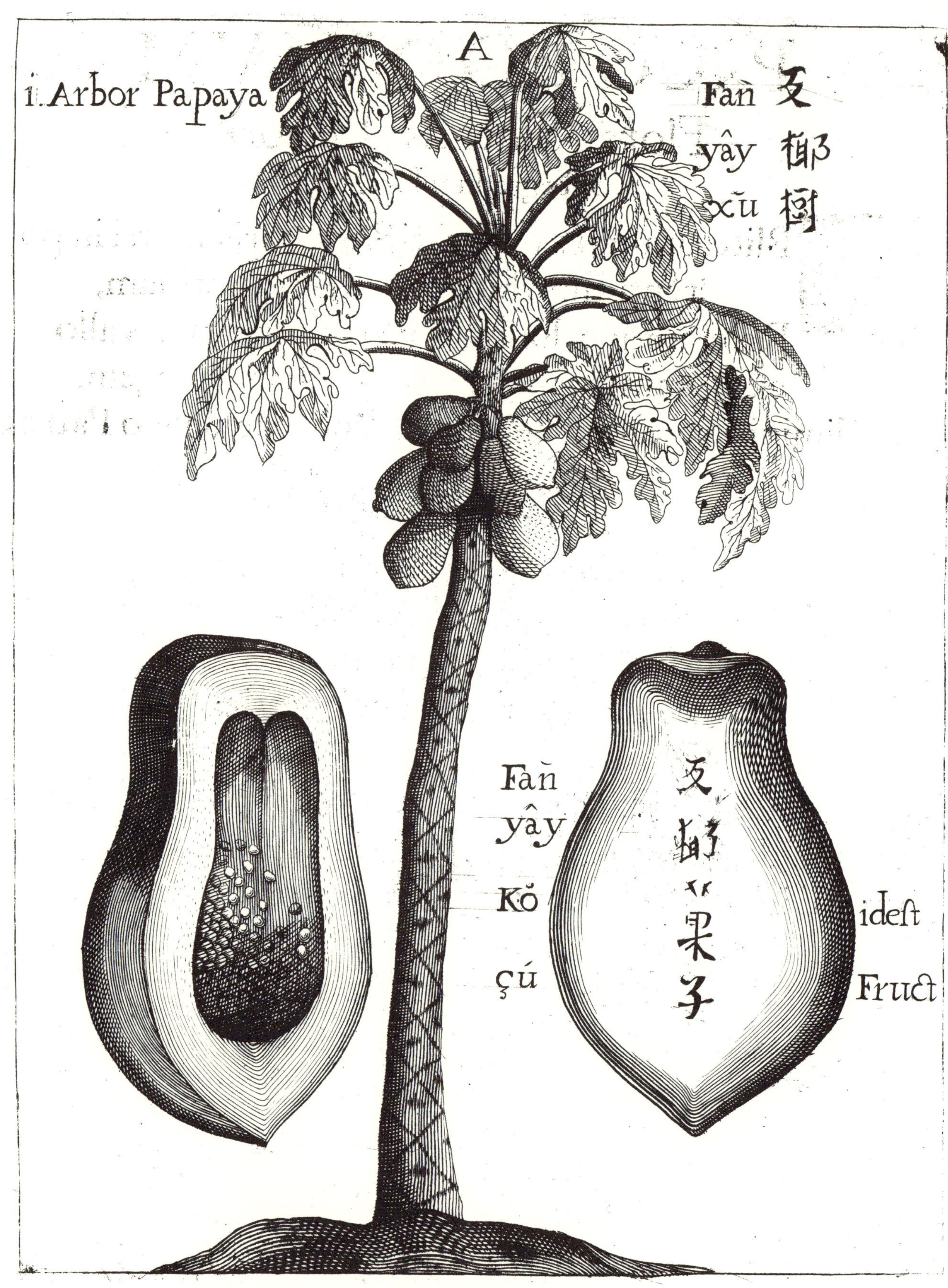

Papaya, *Carica papaya*. From Boym, *Flora sinensis*. Vienna, 1656. (Catalogue item number 46.)

of animals—but, of the eighteen plants depicted in this flora of China, five, the papaya, cashew, pine-apple, guava, and a species of *Annona,* are natives of the new world.

Boym knew that the pineapple was introduced and noted that it had been brought to the East Indies from Brazil, but the other species native to tropical America he apparently thought were native in Hainan and southern China. Some new world food plants, such as peanuts and maize, had been introduced to China early in the sixteenth century, and Ho (1955) recorded the sweet potato from western Yunnan as early as 1563. These are all easily-grown food plants, and the sweet potato, in particular, could be used, and was, as a subsistence food in time of famine. Chili pepper was first reported in 1621 and tobacco three years later.

However, it is surprising to find the pineapple and the four American trees so common in Hainan and the southern provinces of China that Boym seemed to have no doubt that they were native. It is an odd choice of species to include in a flora of China, but Boym admired diversity. He comments that in Europe one sees leaves, flowers, and then fruit appearing on the tree, as is the case with peaches and plums, and, indeed, in China most trees do so as well.

But in China, he says, there are trees which bear flowers on the trunk, or fruit without flowers, and he pictures a species of cauliflorous fig with fruit arising from the roots. Other trees have no branches, but only leaves at the top of the trunk and there produce flowers and fruit—the papaya. He marvels at the tree "ka-giu" which has fruit like apples, but which lack seeds; but at the tip of each fruit is another fruit like a chestnut. This wonder is the cashew with its enlarged fruit stalks. God, Boym says, has produced such a great variety of patterns to show all the possible ways in which plant form can be expressed.

Chabrié, Robert. *Michel Boym: Jésuite Polonais et la fin des Ming en Chine (1646–1662).* Paris: P. Bossuet, 1933.

Ho, Ping-Ti. "The Introduction of American Food Plants into China." *American Anthropologist* 57: 191–201. 1955.

Hu, Shiu Ying. "History of the Introduction of Exotic Elements into Traditional Chinese Medicine." *Journal of the Arnold Arboretum* 71: 487–526. 1990.

Szczesniak, Boleslaw. "The Writings of Michael Boym." *Monumenta Serica* 14: 481–538. 1949–1955.